Botanical Microtechnique

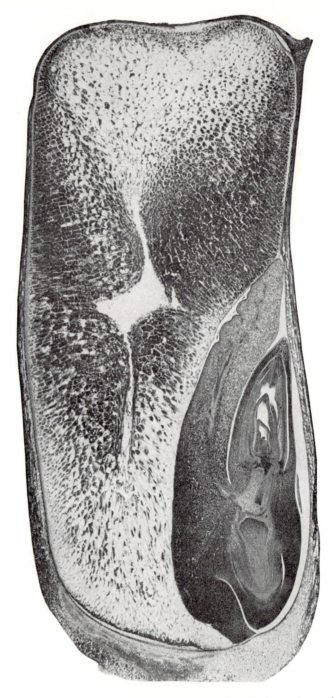

Longitudinal section of kernel of yellow dent maize, 25 days after pollination. Craf III; dioxan-tertiary butyl alcohol; photographed at 8X with B & L 48 mm. Micro Tessar objective; reproduced at 16X.

Botanical
Microtechnique

JOHN E. SASS
Professor Emeritus of Botany
Iowa State University

THIRD EDITION

The Iowa State University Press, *Ames,* Iowa

Second edition, 1951

Third edition, 1958
Second printing, 1961
Third printing, 1964
Fourth printing, 1966
Fifth printing, 1968
Sixth printing, 1971

International Standard Book Number: 0–8138–0220–2
Library of Congress Catalog Card Number: 58-13416

A man who works with his hands is a laborer; a man who works with his hands and his brain is a craftsman; but a man who works with his hands and his brain and his heart is an artist.

—Louis Nizer, *Between You and Me* (Beechhurst)

Preface

Permanent slides for microscopic study are indispensable in the teaching of a basic course in botany and also in specialized advanced courses. In some advanced courses, the students prepare many of the slides used in the course, but in elementary courses the slides are furnished. In the latter case, the slides either are purchased from commercial sources or made in the departmental laboratory. Biological supply houses make excellent slides of the subjects commonly used in elementary teaching, but the quality is likely to be variable. Jobbing houses that purchase slides from constantly changing sources also may furnish disappointing slides at times.

The relative merits of making slides and of purchasing them must be decided on the basis of local conditions. Uncertainties in the commercial supply and the need for specialized or unlisted items necessitates the preparation of slides in the biological departments of schools. This service work often is performed by a skilled professional technician with more or less supervision by the departmental staff. In other departments a member of the teaching staff, usually a morphologist, assumes this responsibility, with the aid of student assistants.

Some research organizations maintain a technician for the preparation of research slides. There are many types of investigation in which it is possible for the technician to prepare and place the finished slides before the investigator, who then carries out the study and interpretation of the material. However, in many investigations, some or all steps in the preparation require an intimate knowledge of the history, structure, and orientation of the material and the aims of the study. The use of a technician who allegedly merely "turns the crank" is then less valid, and the so-called technician may in fact be a research collaborator. The investigator in any field of plant science is urged to utilize microtechnique as a tool, but to do so critically and intelligently and in proper fairness to the workers who contribute their skill, patience, and understanding to the furtherance of the research. It cannot be too strongly emphasized that in order to have a proper appreciation of the possibilities and limitations of present-day techniques, and to utilize the services of commercial or institutional technicians to best advantage, every teacher and investigator in the biological sciences should be familiar with at least the elements of microtechnique. We can do no better than to quote the late Dr. Charles J. Chamberlain, the dean of American microscopists: "The student who has not had sufficient experience to make a first-class preparation for microscopic study cannot safely interpret slides made by others. He is in the same class with the one who claims he sees it but can't draw it; while the real trouble is not in his hand, but in his head."

The term *histology* is very commonly misused to imply histological methods or technique. Histology means the study of the structure and development of tissues, and does not refer to the preparation of slides. A good textbook of histology need not contain a word about sectioning and staining of tissues. A person who takes an afternoon off and learns to whittle some fair freehand sections is neither a histologist nor a technician.

Botanical microtechnique and subsequent histological study may be defined in terms of its functions, which fall into the following overlapping categories:

1. the preparation of plant tissues for microscopic study.
2. the skillful use of the microscope and related equipment for the critical study and interpretation of the material.
3. lucid and concise verbal description and interpretation.
4. recording and illustrating the results by means of the graphic arts.

In some schools microtechnique is taught as part of the work in some branch of morphology, such as anatomy or cytology. That system has marked advantages. The student who has collected and processed his own plant materials, and made his slides, can visualize the orientation of the sections in the plant and interpret the relationship of parts to the whole plant. A disadvantage of the system is that specialized courses in morphology are likely to utilize a limited number of methods — for instance, the smear method in cytogenetics. The student may acquire remarkable skill in making preparations of one type and have no experience with other useful methods. He may develop great skill in making smear preparations of pollen mother cells, but one cannot smear a kernel of corn or a pine stem. He may even acquire disdain for methods which versatile and experienced workers regard as indispendsable.

The maintenance of a separate course in microtechnique makes possible the presentation of the fundamentals of useful standard methods, which experience has shown to be the backbone of research and which have long served the routine needs in teaching. A course should be organized to give a systematic, graded series of exercises, each exercise pointing to some definite objective and yielding superior preparations of a given type. Student interest can be maintained by working with plants that are of interest to the student or the institution, and with plants that are characteristic of the region.

This manual has evolved over a period of years in connection with the teaching of a college course in botanical microtechnique.

Since this is primarily a training manual rather than a reference work, use is made of a graded series of assignments, beginning with subjects in which orientation is easily visualized, few sectioning difficulties are encountered, and a simple stain is used. Subsequent assignments require greater skill in the processing, sectioning, and differential staining of cell and tissue components. A few carefully selected processing and staining methods are presented in detail. Emphasis is placed on gaining an understanding of the aim of the undertaking and the function of every operation, rather than on memorizing and mechanically following a written outline or numbered jars. After mastering the fundamentals, the worker can readily delve into the literature of specialized fields by consulting the brief bibliography. The advanced worker will use the excellent texts by Johansen (1950), Gray (1954), and Jensen (1962), and in particular the comprehensive bibliography in Gray.

The drawings in this book were made by Miss Ruth McDonald. Grateful acknowledgment is made for a grant from the Graduate College for the preparation of the first edition. Thanks are also due to the author's wife, Frances, who did much of the typing on the earlier editions and reprintings. The author's colleagues and students have given much valuable criticism, advice, and encouragement.

JOHN E. SASS

IOWA STATE UNIVERSITY
August, 1958

Contents

Part I
General Principles and Methods

1. Introduction

The study of the microscopic details of the structure of plants usually requires some preparation of the material to facilitate observation. Unicellular, filamentous, or other minute plants require comparatively little preparation. The material may simply be mounted on a slide in a drop of water and thus studied, even under considerable magnification. Larger plants, or parts of plants, must be dissected or cut into thin slices in order to expose inner regions and to permit light to penetrate through the object. Some materials have enough natural coloration to be visible even when finely divided or sectioned. Highly transparent or colorless structures, on the other hand, must be made visible by the use of stains. Preparations that are to receive considerable handling over a period of time should have some degree of permanence. The desirable properties of microscopic preparations are, therefore, adequate thinness, coloration or refractile visibility, and permanence.

The processes used in the preparation of plant materials for microscopic study can be roughly classified in the following categories:

1. Unicellular, filamentous, and thin thalloid forms can be processed *in toto* — without sectioning — and mounted as "whole mounts" to make temporary or permanent slides.

2. Some succulent tissues can be crushed or smeared into a thin layer on a slide or cover glass. The preparation is then stained and treated to make temporary or permanent slides.

3. The more complex and massive tissues are usually sliced into very thin slices, freehand or with a microtome. Materials that are not sufficiently rigid to be cut without support are embedded in a supporting matrix before sectioning. The sections are stained and mounted to make temporary or permanent slides.

The method used for the perparation of a given subject depends on the character of the material, the use that is to be made of the

[3]

slides, and such facilities as equipment, reagents, and time. The experienced worker does not overstress the merits and applicability of some one method. For example, important advances in smear methods and related processes for the study of nuclear and chromosome details have replaced to some extent embedding and sectioning. The whole-mount method is recognized to be entirely satisfactory for many algae, fern prothalli, and similar subjects. However, microtome sections of embedded material must be made if we wish to study the undisturbed cellular organization of a tissue, the development and arrangement of organs, or the structural relationship between a fungus or insect parasite and the tissues of its host. The much-maligned celloidin method must be used to keep intact a badly decayed, fungus-infected piece of oak railroad tie for an examination of the mycelium in the wood. In order to avoid undue emphasis on any particular method, we should recognize that each of the well-established methods has its proper sphere, in which it is the most effective and economical method of performing a given task.

The sequence in which processes are arranged in this book takes cognizance of the fact that the paraffin method furnishes by far the largest number of slides produced for teaching and research. Certain operations, such as the killing of protoplasm and the preservation of fixation images, are essentially similar for smears, sectioning, and whole-mount methods. The preliminary processing of material is much the same in the several embedding and sectioning methods. In view of these facts, the paraffin method is presented with unbroken continuity of its operations.

2. Collecting and Subdividing Plant Materials

The preservation of structural details of cells and tissues is influenced by the condition of the plant at the time of collecting and by the subsequent preparation for killing (fixation). For the study of normal structure, select healthy, representative plants. Remove the plant or the desired part with the least possible injury to the sample. If the material is to be killed at once, follow the procedure outlined in Chap. 3. If the material cannot be killed promptly, it should be stored and transported in such manner that bruising, desiccation, molding, and other injuries are minimized. Do not use material that has been obviously damaged in storage or shipment. The unsatisfactory slides obtained from such material are likely to be interpreted by uncritical observers as the result of poor technique. Dried herbarium specimens can be softened and sectioned to make slides in which it is possible to determine the gross features of vascular arrangement or carpellary organization (Hyland, 1941). However, such material is not suitable for detailed microscopic study.

The following general directions are introduced at this point for the use of readers who have selected subjects on which to work. The reader who seeks suggestions concerning suitable and tested subjects should turn to Part II and use the recommendations made therein in conjunction with the present chapter.

LEAVES

Remove a leaf or leaflet by cutting the petiole, without squeezing or pulling the petiole. The vascular bundles in the petioles of some plants become dislodged easily. For transportation or brief storage, place the leaves between sheets of wet toweling paper and keep in a closed container such as a tin can or a Mason jar. If the leaves appear to be wilted on arrival in the laboratory, freshen them in a moist chamber or in water before processing.

[5]

STEMS

Leafy stems can be kept fresh for several days by standing them in a container of water, preferably in a refrigerator. If such storage is not practicable, cut the stems into the longest pieces that will fit into the available closed container without folding or crushing. Wrap the pieces promptly in wet paper and store in a cool place. Dormant woody twigs, large limbs, and disks cut from logs can be kept for weeks in a refrigerator without appreciable injury.

ROOTS

Do not collect roots or other underground organs by pulling up the plant. The delicate cortex is easily damaged, in fact, the woody stele may be pulled out of the cortex, leaving the cortex in the ground. To collect roots without damaging them, dig up the plant, soak the mass of soil in water until thoroughly softened. Wash the soil away carefully, cut off the desired roots and brush them gently with a camel's hair brush to remove as much soil as possible. Wrap the pieces and store as in the case of stems.

FLORAL ORGANS

Remove entire flowers or flower clusters and wrap in wet paper. Store in a closed container in a cool place. Large buds like those of lily can be kept in a Mason jar of water until you are ready to dissect and preserve them. Fruits may be collected and stored in a similar manner.

LIVERWORTS AND MOSSES

Remove groups or mats of the material with a generous quantity of the substratum. Store in a moist chamber until the plants are turgid. Saturate the substratum in order to permit the removal of complete plants without excessive damage. Dissect out the desired parts under a binocular and transfer to the preserving fluid promptly.

ALGAE

Collect in a quantity of the water in which the plants are growing, and keep in a cool place in subdued light. Many filamentous forms disintegrate rapidly in the laboratory, and even in the greenhouse unless the temperature and light can be carefully controlled. It is best to kill algae promptly after collecting.

FLESHY FUNGI

The larger fleshy fungi can be transported and stored, wrapped loosely in waxed paper. Sporulation continues and may indeed be promoted in this manner. However, since molding and disintegration take place during prolonged storage, material should be processed promptly. Small fungi should be wrapped in moist paper, enclosed in waxed paper, and processed as soon as possible.

PATHOLOGICAL MATERIAL

Particular care should be exercised to insure that the condition of the host tissues is not altered by handling, in order that abnormal structure may be properly interpreted as an histological symptom of the disease. Prevent wilting of the material, or revive it in a moist chamber, but avoid the development of bacteria, molds, or other secondary organisms. For a pathological investigation, always collect normal, disease-free tissues of age comparable with the diseased samples. It is imperative to work out the best technique for preserving the "normal" condition of the host before attempting an authoritative interpretation of slides of pathological material.

The foregoing general remarks will serve as a basis from which the worker can develop effective methods and habits of collecting and handling material in accordance with facilities and circumstances. Hold rigidly to the view that the finished slide should represent the original structure of the plant, whether that structure is presumably normal or pathological or is the result of experimental treatment.

The handling of materials that are to be used for bulk specimens or whole mounts is described in Chap. 10. The preparation of permanent slides from microtome sections consists essentially of the following processes:

1. Selecting desired plants or parts of plants and, if necessary, subdividing into suitable pieces.
2. The killing and preservation of the contents of cells and the preservation of cellular structures in a condition approximating that in the living plant.
3. Embedding in a matrix if necessary, in order to support the tissues for sectioning. See page 91 for the sectioning of unembedded tissues.
4. Sectioning of the tissues into very thin slices.
5. Staining the slices and covering with a cemented cover glass to make a permanent slide.

Subdividing Material for Processing

Some preliminary remarks concerning the action of reagents in the preservation of cells and tissues will aid in understanding the following description of this process. The reagents used for killing

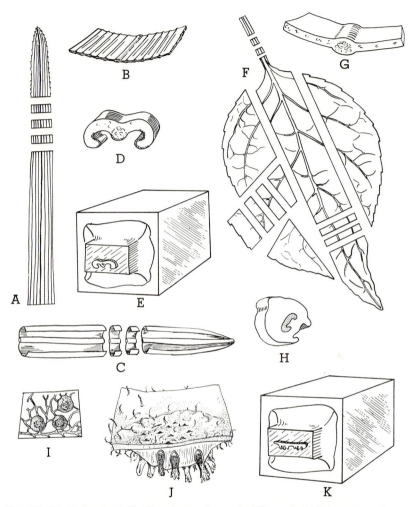

Fig. 2.1—Methods of subdividing leaves for embedding: *A–D,* long narrow leaves and transverse pieces removed from such leaves; *E,* embedded piece of leaf fastened to mounting block; *F–H,* large broad leaf and excised pieces of blade and petiole; *I,* portion of leaf with fungus pustules; *J,* enlarged view of excised aecia; *K,* embedded piece of leaf bearing aecia, fastened to mounting block.

cells contain ingredients that are toxic to protoplasm. In order to stop life processes quickly and without distortion of structure, the killing fluid must reach the innermost cells of a piece of tissue before disintegration takes place. Most reagents penetrate very slowly through the cuticle or cork on the surfaces of plant organs, but penetration is much more rapid through cut surfaces. Therefore, it is desirable to cut the organ being studied into the smallest pieces that will show the necessary relationship of parts.

The subdividing of soft fresh material is best done with a razor blade, with the material placed on a sheet of wet blotting paper or held carefully against a finger. Excessive pressure against the support is likely to rupture delicate tissues as in the mesophyll of leaves (Fig. 11.1) or the chlorenchyma of a stem (Fig. 11.2). Such damage does not become visible until the sections in the ribbon are examined or possibly not until the finished slide is examined. The usual results are peeling of the epidermis and distortion of the crushed tissues.

Leaves are almost invariably cut into small pieces for processing. Narrow leaves that are not much over 5 mm. wide, may be cut into complete transverse pieces measuring 2 to 4 mm. along the rib (Fig. 2.1 *A–D*). Examples of this type are bluegrass, garden pinks, hedge mustard, and some narrow-leaved milkweeds. Broad leaves should be cut into small pieces, selected to include midrib, lateral veins, fungus pustules, fern sori, or other desired structures (Fig. 2.1 *F, G, I, J*). The enlarged views of the pieces of leaf (Fig. 2.1 *B, D, G, I, J*) and the pieces of embedded tissue mounted on blocks ready for sectioning (*E* and *K*) will aid in visualizing the orientation of pieces. Particular care should be used in subdividing pathological material (Fig. 2.1 *I, J*). If it is necessary to know which is the long axis of the leaf, cut all pieces so that the shorter dimension is along the long axis of the leaf, or vice versa, and record the method in your notes.

Herbaceous stems, roots, petioles, and other more or less cylindrical organs are usually cut into short sections or disks. When cutting out sections or subdividing pieces, do not roll or press the pieces. Keep the material moist, and work rapidly. After the final subdivision, drop the pieces into the killing fluid promptly. By means of descriptions and sketches, like those in Figs. 2.1 and 2.2, keep an accurate record of the part of the plant from which the pieces of tissue were obtained.

Figure 2.2 gives additional suggestions for subdividing organs. A stem that does not exceed 2 mm. in diameter should be cut into sections 2 mm. long if highly cutinized, but may be as long as 10 mm. if the surface is permeable. An organ 5 mm. in diameter should be cut into 5-mm. lengths. An organ 1 cm. in diameter should be cut into disks 2 to 5 mm. thick. Stems of larger diameter are usually cut into 5-mm. disks that are halved or quartered longitudinally or divided into wedge-shaped pieces.

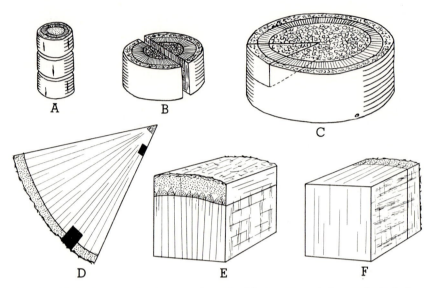

FIG. 2.2—Methods of subdividing massive cylindrical organs: *A–C*, sample includes portions of all tissues in the axis; *D* shows the position of pieces removed from a large log; *E* and *F*, enlarged views of trimmed pieces removed from large log.

Woody twigs having a diameter up to 5 mm. should be cut into 15-mm. lengths. Larger twigs should be cut into shorter pieces because the impermeable cork makes penetration by reagents difficult, except through the cut ends. Do not cut the twigs into pieces with pruning shears or a knife. Rough handling will bruise the cambium, phloem, the fragile primary cortex and cork cambium, resulting in the separation of the outer layers during sectioning or during staining. Use a razor blade and cut through the twig by chipping a groove deeper and deeper around the twig until it is cut through. An excellent tool for cutting twigs into sections is a fine-toothed high-speed saw, such as a rotary dental saw or a jig saw, especially the vibrating diaphragm type.

To make slides of transverse, radial, and tangential sections in the region of the cambium of old trees, use tissues removed from newly felled logs or limbs having a diameter of at least 10 cm. Cut disks 2 to 3 cm. thick from portions of the log that were not bruised in felling. Wrap the disks in wet burlap and take into the laboratory at once for further trimming. Split a disk radially into pieces having uninjured blocks of bark firmly attached to the wood. Trim off enough of the inner part of the wedge of wood to leave a block of sapwood with several annual rings and with cambium and all outer tissues intact (Fig. 2.2 *D-F*). With a razor blade split a thin layer from the two radial faces, from the inner tangential surface and from the transverse faces of the block, thereby removing tissues that were compressed during the preliminary trimming. Keep the material wet during these operations. Drop the pieces into the killing fluid at once after final trimming.

Wood from dead logs, dry lumber, or furniture wood requires proper trimming to establish the future cutting planes. It is usually easy to establish the radial plane by splitting the wood longitudinally, parallel to a ray. At right angles to this plane, split the block longitudinally along the tangential plane, and then trim in the third or transverse plane. Rough splitting can be done best with a plane bit, and rough crosscutting with a fine-toothed high-speed mechanical saw. Finally, trim all faces with a razor blade to remove surface tissues that were damaged by the rough trimming. Subsequent processing of the wood is described in the section on the preparation of hard tissues.

The handling of more specialized and difficult materials such as buds, floral organs, and fruits is described to better advantage in Part II in conjunction with detailed directions for processing such materials. The handling of plant bodies and organs of the lower phyla is also described in Part II.

The foregoing brief outline of methods of collecting and preparing material for preservation can be modified and adapted to meet most problems. The principal preliminary operations and precautions necessary for successful processing may be summarized as follows:

1. Use fresh, normal material.

2. Remove pieces having the desired features and oriented so as to establish planes in which microtome sections are to be cut.

3. Cut into suitable pieces, with minimum bruising, compression, or desiccation.

4. Immerse the pieces promptly into the killing (fixing) fluid (Chap. 3), and promote quick penetration of the fluid by removing air with an aspirator (Fig. 3.1).

5. Record the necessary data concerning species, location, date, parts selected, and killing fluid used.

3. Killing, Fixing, and Storing Plant Tissues

One of the most critical operations in the processing of tissues is the killing of the protoplasm. The stopping of life processes within the cells should be accomplished with the minimum structural disturbance within the cells and minimum distortion of the arrangement of cells in the tissues. In addition to killing the protoplasm, the killing fluid or the subsequent processing must retain or fix the undistorted structure and render the mass of material firm enough to withstand the necessary handling.

No single substance has been found to meet the requirements of successful preservation. The formulas used for this purpose consist of ingredients in such proportions that there is a balance between the respective shrinking and swelling actions of the ingredients. The numerous formulas found in the literature are variations of a comparatively few fundamental formulas, and the chemical substances in the formulas are few in number. Any formula should be regarded as a starting point for experiments to determine the proper balance of ingredients for specific subjects. The formulas recommended in this chapter have been found to be satisfactory for a diversity of subjects.

Preparation of Stock Solutions and Killing Formulas

The following reagents and stock solutions are used in a wide range of killing (fixing) formulas:

Glacial acetic acid.
　1% acetic acid (approximately) , made by adding 10 cc. of glacial acetic acid to 990 cc. of water.
　10% acetic acid, made on the same basis as the above.
Propionic acid may be substituted for acetic acid in the above.
1% chromic acid. (10 g. chromic anhydride crystals per liter.)
Formalin, the trade name used for an aqueous solution of formaldehyde, containing 37 to 40% formaldehyde gas by weight.

Picric acid, saturated aqueous solution.

2% osmic acid. 2 g. crystals in 100 cc. of 1% chromic acid, or in 100 cc. of water.

Ethyl alcohol; commercial 95% grade and anhydrous grade.

The use of stock solutions of 1% and 10% acetic or propionic acid is advocated because the error involved in measuring a small volume, say 1 cc., of glacial acetic acid is much greater than in measuring 10 cc. of 10% acid.

Apparatus

Use specimen bottles that hold a generous quantity of killing fluid, especially with bulky or succulent materials that may dilute the formula. After washing and partial dehydration, materials may be transferred to smaller bottles or vials for the remainder of the process.

When the pieces of plant material are dropped into the killing fluid, the hairs, stomates, folds, and other cavities of plant organs retain air bubbles which retard penetration by reagents. If the pieces do not sink at once, attach the bottle to an aspirator, and apply suction for repeated short intervals until the pieces sink, if not to the bottom of the liquid, at least under the surface. Use a safety bottle (Fig. 3.1 *A*) to keep water from backing into the specimen

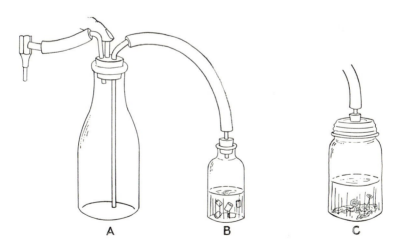

A   B   C

FIG. 3.1—Aspirator setup for pumping specimens in killing fluid: *A*, safety bottle with finger valve or glass stop clock; *B*, specimen bottle or large empty bottle into which specimen bottle is placed; *C*, pint jar used as container for large specimens.

bottle. Tapping the specimen bottle gently against the sink aids in the loosening of air bubbles within or on the specimen. Highly buoyant materials should be placed into a tall vial of the killing fluid and held below the surface by means of a plug of cheesecloth. A screw-topped wide-mouthed bottle is necessary for evacuating large objects (Fig. 3.1 *C*). When most of the pieces remain submerged after the suction is released, push any floating pieces under the surface with a matchstick, and most of them will then sink. Remove and discard all pieces that do not sink after pumping and submersion.

Materials from which it is difficult to evacuate air do not become infiltrated readily and should be pumped again when nearing the end of the dehydrating series, and again when in the final change of paraffin solvent, before any paraffin has been added. Connect a second safety bottle between the regular safety bottle and the specimen. The possible entry of water vapor into the specimen bottle when the pump is shut off is prevented by having a deep layer of calcium chloride and a layer of cotton in the second safety bottle. The ingenious and precisely controllable vacuum apparatus of Wittlake (1942) may be used for the killing, as well as the subsequent operations of embedding.

Killing and Fixing of Tissues

Killing solutions may be grouped into types on the basis of the ingredients used. Some formulas are stable and may be kept on hand ready for immediate use. Other formulas must be made up immediately before use. The formulas given on the following pages have been computed so that they can be made up from the above stock solutions by volumetric measurements. The system of letters and numbers used in this manual to designate killing fluids is explained later in this chapter.

The length of time necessary to bring about killing and hardening of material varies greatly and is determined by the character of the fluid used, the bulk of the individual pieces, and the resistance of materials to penetration by reagents. Fluids of the anhydrous type, such as Carnoy's absolute alcohol-glacial acetic acid formula, penetrate small objects almost instantaneously, and killing and hardening are a matter of minutes. The chrome-acetic fluids penetrate slowly into the interior of a piece of tissue, and have poor hardening action. Recommendations concerning the duration in killing fluids are given in the description of the various formulas. Washing of tissues, which is necessary after some killing fluids, is discussed in connection with specific formulas.

One of the most useful types of killing and preserving fluid, known as *FAA,* is represented by the following formula:

Ethyl alcohol (95%) .... 50 cc.
Glacial acetic acid .... 5 cc.
Formaldehyde (37–40%) .... 10 cc.
Water .... 35 cc.

Propionic acid may also be used, the formula is then designated *FPA.*

Several modifications may be found in the literature. This fluid is stable, has good hardening action, and material may be stored in it for years. These properties make this formula suitable for large or impervious objects such as woody twigs, tough herbaceous stems, and old roots. The high concentration of alcohol is likely to produce shrinkage of succulent materials, although it is possible to develop a formula for some apparently tender subjects and even for filamentous algae. A balanced formula can be worked out by varying the acetic acid, which has a swelling action on protoplasm, from 2 to 6% by volume. The formaldehyde and alcohol, which have a shrinking action, should be held at the indicated concentrations. When making trials of variations from the fundamental formulas, kill a trial lot or batch of material in the formula to be tested, and a check lot in a standard formula, and carry the batches through identical processing simultaneously, so that differences in cellular detail will be the result of variations of formula.

Pieces of thin leaf are killed and hardened in 12 hr. The actual killing of the protoplasm probably occurs in much less time. Thick leaves or pieces of small stem require at least 24 hr. Woody twigs should be kept in *FAA* at least a week before continuing the processing for embedding. Materials do not need to be washed after *FAA.* The ingredients of this fluid are soluble in the dehydrating agents and are thus removed before infiltration is begun.

An extensively used formula consists of *FAA* containing bichloride of mercury ($HgCl_2$) to saturation. This fluid penetrates and hardens tissues rapidly. It preserves bacterial zoogloea in plant tissues, thus being useful in pathological studies. The alcohol may be increased to 70%. Prolonged storage in fluids containing bichloride is undesirable. The tissues should be transferred after 48 hr., or at most a week, to a fresh solution of the original formula which does not

TABLE 3.1

KILLING FLUIDS OF THE CHROME-ACETIC AND FLEMMING TYPE*

(The numbers in the columns represent cubic centimeters of the designated reagents)

Stock solution	Chrome-acetic					Flemming type			
	Weak I	Weak II	Medium I	Medium II	Strong	Weak	Medium	Strong	Chamberlain
1% chromic acid.	30	50	50	70	97	25	50	75	96
1% acetic acid...	70	50	.....	.....	.....	10	.....	.....	.....
10% acetic acid..	.....	.....	10	20	.....	.....	10	.....	.....
Glacial acetic acid	.....	.....	.....	.....	3	.....	.....	5	3
2% osmic acid...	.....	.....	.....	.....	.....	10	10	20	1
Water..........	.....	.....	40	10	.....	55	30	.....	.....

* The formulas in Tables 3.1 and 3.2 have been arranged and arbitrarily numbered, beginning with the weaker solutions.

contain the bichloride of mercury. After four or five changes of the latter solution, tissues may be stored indefinitely in the last change.

Chromic acid and acetic acid are the ingredients of an important class of fluids, the *chrome-acetic* formulas. These fluids are not used as extensively as some years ago. Table 3.1 gives the proportions of five formulas. Because of widespread use of the terms weak, medium, and strong for fluids of the type given in Table 3.2, these terms are retained for the series of modifications in the table. The weaker solutions are suitable for succulent or delicate subjects, the strong solution for firm subjects. If this type of fluid is to be used for a critical study, a balanced formula should be worked out by balancing the shrinking action of chromic acid and the swelling action of acetic acid.

The above formulas are not very satisfactory for bulky or woody subjects because of poor penetrating ability. Use these mixtures for filamentous and thalloid plants, root tips, floral organs, and small sections of leaves or stems. The length of time required to kill materials varies greatly. Filamentous algae are probably killed in a few minutes. Small pieces of leaf or root tips require about 12 hr. Larger pieces of tissue should have at least 24 hr. The progressive destruction of chlorophyll from the cut edges inward is a good gauge of the rate of action. Prolonged storage in chrome-acetic produces brittleness of the tissues and muddy staining effects. Therefore, these fluids are not suitable for storage and the tissues must be processed after the optimum interval necessary for killing.

Materials killed in the fluids given in Table 3.1 should be washed in running water. Various devices may be used for accomplishing this prolonged washing. The simplest method is to tie a strip of cheesecloth over the wide mouth of the bottle containing the tissues and to allow a slow stream of water to flow into the bottle. More vigorous washing action can be obtained by inserting the water inlet tube to the bottom of the specimen bottle. These fluids do not have good hardening action, so it is best to avoid violent motion of the pieces. Firm materials can be washed in a vertical length of 1-in. glass tube with a stopper at the lower end, admitting a stream of water through a small tube, the waste water leaving through cheesecloth tied over the upper end of the large tube.

Osmic acid is used in a class of formulas known as the Flemming fluids. These fluids are indispensable for cytological studies but are seldom justifiable for histological work. Osmic acid is expensive, its vapors are highly irritating, and it blackens tissues, making it necessary to bleach sections before staining. Osmic acid preserves chromosome details with great fidelity, but has no special virtues for the preparation of slides of such subjects as corn stem or apple leaf for anatomical or histological study. Osmic acid has poor penetrating ability and is therefore not satisfactory for bulky objects. The formulas given in Table 3.1 will serve for preliminary tests, subject to experimental variation of proportions. Because of the blackening action and poor hardening properties of the Flemming fluids, material should be washed in water and processed immediately after killing. The intervals for killing are approximately those given for chrome-acetic.

Table 3.2 gives several formulas based on the Nawaschin formula, containing chromic acid, acetic acid, and formaldehyde. Numerous modifications may be found in the literature. The name Craf has been coined for this widely used type of fluid. For critical work on specific subjects, experiment with variations of the formulas in the table. The acetic acid should be varied from 0.7 to 5% glacial acetic acid equivalent by volume. The optimum chromic acid and formaldehyde concentrations for many subjects are the proportions given in formula V. The other formulas in the table, including Nawaschin's original formula, also give good results with specific subjects. The formaldehyde should be added immediately before using. If one of these formulas is to be used for making extensive collections in the field, it will be found convenient to make up the desired mixture of the chromic and acetic acids, adding the measured volume of for-

TABLE 3.2

KILLING FLUIDS BASED ON THE NAWASCHIN AND BOUIN FORMULAS

(The numbers in the columns represent cubic centimeters of the designated reagents)

Stock solution	Nawaschin type (Craf)						Bouin	Allen-Bouin type		
	Nawaschin	I	II	III	IV	V		I	II	III
1% chromic acid..	75	20	20	30	40	50	......	50	50	25
1% acetic acid....	....	75	....	....	....	....	......	....	....	....
10% acetic acid...	....	....	10	20	30	35	......	20	....	40
Glacial acetic acid.	5	...	...	...	...	...	5	....	5	...
Formaldehyde 37–40% aqueous...	20	5	5	10	10	15	25	10	10	10
Picric acid saturated aqueous....	....	....	....	....	....	....	75	20	35	25
Water...........	....	....	65	40	20	....	......	....	....	....

maldehyde before using. A few hours after the formaldehyde is added a perceptible change of color takes place in the liquid, and after several days the chromic acid becomes changed to an olive or green compound. Long before this condition is reached, killing action has been completed, and the altered fluid then serves as an excellent hardening and preserving agent. Material may be left in these fluids for as long as 5 years and yield excellent histological preparations. The effect of prolonged storage on critical cytological details deserves further study. The minimum time for small masses of soft tissue is 12 hr., but it is obvious from the foregoing remarks that several days at least can be allowed to insure hardening without danger of distortion or darkening of the material. A further advantage of the Nawaschin type fluid is that materials need not be subsequently washed in water, thus avoiding the possible softening and pulping of material.

Bouin's fluid, given in Table 3.2, has long and deservedly been a favorite. It is excellent for root tips, especially for telophase figures, and has been used successfully for embryo sac studies. The complete mixture is stable and may be kept on hand in the laboratory or carried to the field ready for use. A minimum interval of 12 hr. is suggested for finely divided material. Larger pieces such as thick root tips or mature tissues should have at least 48 hr. Prolonged storage is regarded as undesirable. After the optimum interval in the killing fluid, the material is not washed in water, but is rinsed several times in 20% alcohol or acetone. Dehydration is then continued as described later.

The addition of chromic acid and urea to Bouin's fluid makes what is known as the Allen-Bouin formula. For cytological work use the original formula, as given in the reference manuals, or one of the formulas (lacking urea) given in Table 3.2. For further trials vary the glacial acetic acid equivalent from 1 to 4% by volume. The formaldehyde should be added immediately before using. Tests have shown that tissues may be left in these solutions for several months. It is probable that hardening of the material reaches a maximum in less than a week. Dehydration and subsequent processing are carried out as with Craf.

Farmer's fluid and Carnoy's fluid have limited uses in histology. These fluids kill protoplasm by rapid and probably violent dehydration. Because of their ability to penetrate very rapidly, these fluids have some value for processing extremely downy, resinous, or impermeable structures that must be preserved entire. The fluid may be used alone, followed in 1 hr. or less by the subsequent operations of the paraffin process. An alternative method consists of first immersing the materials in a Carnoy or Farmer formula (the time ranging from an instantaneous dip to 10 min.) and then treating in one of the more critical fluids. Two widely used formulas are as follows:

1. Farmer's fluid
   Anhydrous ethyl alcohol .... 75 cc.
   Glacial acetic acid .... 25 cc.
2. Carnoy's fluid
   Anhydrous ethyl alcohol .... 60 cc.
   Glacial acetic acid .... 10 cc.
   Chloroform .... 30 cc.

The fluids given thus far in this chapter produce an *acid fixation image,* preserving particularly well the chromosomes, nucleoli, and the spindle mechanism. Nucleoplasm and mitochondria are dissolved; cytoplasm is rendered in fibrillar or alveolar form. This type of image is preferred for most studies of plant structure.

In certain cytological studies it is desirable to preserve mitochondria and allied cytoplasmic structures. In such cases a fixing fluid that produces a *basic fixation image* is used. Such fluids preserve mitochondria, nucleoplasm, and in some instances nucleoli and vacuoles. Chromatin and the spindle mechanism are dissolved. For serious studies in this field of cytology each worker must work out specific techniques based on an extensive literature. However, it is possible to produce slides showing mitochondria adequately for teaching purposes, using Zirkle's modification of Erliki's fluid.

Water	400 cc.
Potassium bichromate	2.5 g.
Ammonium bichromate	2.5 g.
Cupric sulphate	2.0 g.

Fix for 24 to 48 hr., wash in water, dehydrate and embed in paraffin.

The desirability of wetting agents and penetrants in microtechnique has been apparent to experienced workers for many years. The rapid development of numerous wetting agents in recent years has led to considerable experimentation, with the expected diverse results. The most prominent unfavorable effect of wetting agents are the peeling of cuticle and epidermis, and varying degrees of cell distortion. Further experimentation with the increasing number of available substances is certainly desirable.

The wetting action of a substance can be tested easily. Make a series of solutions of the substance to be tested by diluting a 1:1000 stock solution. Cut uniform pieces of a highly pubescent leaf and drop alternating pieces into distilled water and into the dilutions of the wetting agent. Note the relative time required for the leaf pieces to sink, and use the most dilute wetting agent that will sink the tissues after brief aspirating. Determine whether the wetting agent forms a precipitate or cloudiness with the killing fluid. If a reaction occurs, do not add the wetting agent to the fixing fluid, but sink the tissues in the wetting agent, rinse with water and cover with the killing fluid. The final criterion is the condition of the tissues in the finished slide, compared with tissues processed without the wetting agent.

A workable terminology for designating killing fluid formulas is a great convenience for giving oral or written instructions, or making routine records. The name of the investigator who first devised a type of formula is not always a satisfactory designation because the proportions of the ingredients are necessarily varied for different subjects. An arbitrary number is not sufficiently descriptive, except among a group of closely associated workers. The terminology proposed here is a compromise, the type of formula is indicated by a name or abbreviation, and the proportion of ingredients by a percentage figure. The proportion of a solid like chromic acid is given as precentage by weight; liquids like melted glacial acetic acid are given as percentage by volume. For instance, the time-honored chrome-acetic has numerous variants, one of which is C-A 0.5-0.5, meaning 0.5% chromic acid by weight, and 0.5% acetic acid by volume. Table 3.1 gives the proportions of stock solutions used to make 100 cc. of mixtures in this category.

A variant of the Nawaschin formula, Craf 0.20-1.0-10.0, contains 0.2% chromic acid, 1.0% acetic acid, and 10.0% commercial form-aldehyde solution. A variant of the Allen-Bouin formula is designated A-B 0.20-4.0-10.0-25.0, containing in addition to the ingredients of Craf, a saturated aqueous solution of picric acid, 25.0% by volume.

The foregoing system of terminology is accurate, descriptive, and convenient and has been used successfully by beginners and advanced workers.

4. Dehydration for Embedding

This operation removes water from the fixed and hardened tissues. Dehydration has some washing action, and makes the material firm and possibly hard and brittle. The process consists of treating the tissues with a series of solutions containing progressively increasing concentrations of the dehydrating agent and decreasing concentrations of water. Two contrasting methods are used to dehydrate and prepare materials for infiltration. In the first method to be described, the tissues are dehydrated in a nonsolvent of paraffin and then are transferred to a solvent. In the second method, the dehydrant is also a solvent of paraffin. The first method of dehydration also is used prior to infiltration in celloidin.

Dehydration by Nonsolvents of Paraffin

The most commonly used dehydrating agent in this category is ethyl alcohol. This is usually purchased in two grades, commercial 95% grain alcohol and absolute (anhydrous) alcohol. The solutions in the dehydrating series are made by diluting 95% alcohol with distilled water. After ascertaining the exact concentration of the alcohol purchased from a given source, it is easy to compute a table giving the respective proportions of alcohol and water for each solution in the series. However, since the series is intended to consist of a graded series of solutions rather than definite concentrations, it is quite adequate to assume the 95% commercial alcohol to be 100% and make up a series containing (approximately) 5, 10, 15, 20, 25, 30, 35, 40, 45, 50, 60, 70, 80% alcohol by volume. Next in the series is the undiluted commercial alcohol (actual 95%), followed by anhydrous alcohol. This graded series of solutions should be kept on hand in the laboratory.

As discussed in the preceding chapter, some killing fluids require more or less prolonged washing of the tissues in water; other fluids require no washing, and dehydration is begun directly after killing or after a brief rinsing in water. Begin dehydration with a dehydrant having approximately the same percentage of water as the killing or storage fluid. For example, after *FAA,* begin dehydration in 50% alcohol. After weak chrome-acetic or the weaker Craf type formula, such as I and II, begin in 5 or 10% alcohol. The stronger formulas such as III, IV, and V, which have greater hardening action, make it possible to begin dehydration in 20 or 30% alcohol. When Craf V was in use for routine chromosome counts, root tips were transferred directly from the killing fluid into 75% alcohol. After Bouin's fluid, begin with 50% for firm subjects and 20% for delicate materials.

Solutions in the dehydrating series are changed by decanting the liquid from the tissues and promptly flooding the material with a generous volume of the solution next in the series. A piece of fine brass-wire screen or a layer of cheesecloth is used to retain materials that tend to float out of the bottle. The volatility of the solutions high in the series demands speed in making the change to avoid drying of the tissues. The material should not be permitted to become dry even for an instant at any stage in the process. Never drain the fluid from several specimen bottles, and then look on the shelf for the next reagent, only to find that the bottle is empty.

The interval in each of the solutions in the series depends on the size of the pieces, the nature of the material, and the solubility of the residual reagents left in the tissues. For root tips or small pieces of leaf use 30-min. intervals up to 70%. After a picric acid formula make each interval 1 hr. For twigs killed in *FAA* use 4- to 8-hr. intervals up to 70%. For large blocks of wood the interval should be about 12 hr. Beginning with 70%, double the previous interval for each grade. Change the cork for a thoroughly dry one when first changing to 100% alcohol. Make three changes of anhydrous alcohol. Plan the timing of the dehydration series so that the series is stopped at 70% for storage until you can resume the process.

Some workers are inclined to make an unnecessary ritual of the time element in dehydration. It is recognized that drastic changes of concentration bring about shrinkage of protoplasm and distortion of cells. Long intervals in low concentrations of dehydrating fluid, or long washing in water, tend to make tissues soft and promote disorganization. Long exposure to high concentrations or anhydrous reagents shrinks tissues and causes brittleness. With these general precautions in mind the intervals can be regarded as sufficiently flexible to conform to the demands of other duties.

Isopropyl alcohol can be used in exactly the same manner as ethyl alcohol. Isopropyl alcohol can be purchased without restrictions, and the commercial grade can be dehydrated as described on page 29. Methyl alcohol has not been used extensively for dehydrating plant tissues. Its toxicity is objectionable, and the vigorous dehydrating action damages delicate structures.

Acetone is an excellent dehydrant. Its purchase and use present no legal, administrative, or disciplinary problems, making it a desirable substitute for ethyl alcohol. Acetone is obtainable in several grades, at prices that vary widely with the quality and source. If anhydrous acetone can be purchased in drum lots at reasonable cost, only this one grade needs to be stocked and used for all the dehydrating grades. Acetone of good quality, but not strictly water-free, can be obtained and used for the gradations, and the more expensive anhydrous grade used only for the final stages in the process. The procedure with acetone is exactly the same as with ethyl alcohol. It is permissible to change from alcohol, or a killing fluid containing alcohol, to a grade of acetone having approximately the same water concentration.

Acetone is highly volatile, and care should be taken not to permit acetone to evaporate from tissues or slides during processing.

Glycerin is used as a dehydrant, especially for algae and other delicate subjects. The high boiling point of glycerin permits the elimination of water by evaporation. The slow, progressive dehydration prevents sudden changes of concentration and minimizes plasmolysis. Material must be washed in water before using glycerin, because the evaporation process obviously does not wash residual reagents out of the tissues. Moderately firm tissues can be washed in running water, but delicate materials should be washed by diffusion. Rinse the material carefully to remove the bulk of the killing fluid, transfer to a 2-quart jar of water, and allow the jar to stand undisturbed for 2 hr. Siphon off most of the water without agitating the material, and refill the jar with water. Repeat the replacement of water at least twice, then proceed with the glycerin method.

Transfer the material to a large volume of a 5% solution of glycerin in water. Use a wide-mouthed bottle or jar and mark the level of the 5% glycerin. The volume should be so gauged that after the elimination of water the residual glycerin will more than cover the tissues. Evaporation of water may be accomplished by several methods or combinations of methods. The most practicable are as follows:

1. In an incubator oven at 35 to 40°C.
2. In a desiccator at room temperatures or in the above oven.
3. In a vacuum desiccator or vacuum oven.

If the glycerin solution becomes colored or turbid during evaporation it may be replaced with fresh glycerin solution of the same concentration. When the volume of the solution has been reduced by evaporation to one-half of the original volume, the glycerin concentration is approximately 10%, and the liquid may be replaced with fresh 10% glycerin and the evaporation continued. Most of the water can be removed by evaporation, especially in vacuum. After a nearly anhydrous condition is attained, the tissues are firm enough to withstand transfer directly into anhydrous alcohol. Change the alcohol at least twice, and proceed with the graded transfer to the desired paraffin solvent as described below, or proceed with one of the whole-mount methods (Chap. 10).

TRANSFER TO A SOLVENT OF PARAFFIN (CLEARING)

After the use of dehydrating agents that are not solvents of paraffin, the dehydrated tissues are transferred to a solvent. The term *clearing,* applied to this transfer, is derived from the fact that some paraffin solvents render the tissues transparent. The clearing action is merely incidental to the function of the reagent, to serve as a solvent of paraffin. The most common solvents are xylene (xylol) and chloroform. Either reagent may be objectionable or even toxic to some workers. Xylene is inexpensive and is by far the most widely used solvent. Chloroform is more expensive, but it is less likely to be toxic. Benzene and toluene can be used, but their lower boiling points increase the fire hazard.

As in the case of dehydration, a graded series is used for clearing. After dehydration in ethyl alcohol, the following absolute alcohol-xylene series is used. For critical cytological work 10 gradations have been recommended. The interval in each mixture ranges from ½ hr.

Grade number	Ethyl alcohol %	Xylene %
1	75	25
2	50	50
3	25	75
4	0	100

for very small or thin pieces to 3 hr. or more for large pieces of tissue. A similar acetone-xylene series can be used.

Chloroform may be substituted for xylene in a similar series, except that more abrupt changes are permissible. A practical series is as follows:

(1) ⅓ chloroform
     ⅔ absolute alcohol
(2) ⅔ chloroform
     ⅓ absolute alcohol
(3) Pure chloroform, changed at least once

Chloroform does not make tissues as brittle as does xylene.

Trichloroethylene is a good solvent of paraffin and may be substituted for xylene in the foregoing processes. Trichloroethylene is not inflammable and is not toxic unless inhaled directly in large quantities. It dissolves Canada balsam but does not affect stained sections. This reagent deserves thorough trial with a wide range of subjects. Any reagent that decreases the hazards of fire and poisoning is worth serious consideration.

Cedar oil is an excellent clearing agent after dehydration in ethyl alcohol. The procedure is to pour a layer of cedar oil into a dry vial, then carefully pour the anhydrous alcohol containing the material over the cedar oil. The pieces gradually sink into the oil and become strikingly clear. The alcohol is removed with a pipette, and the cedar oil is rinsed out of the tissues with several changes of xylene.

Recognition of the fact that the transparency of the tissues at this stage of the process is of no value, and the widespread use of the higher alcohols for dehydration and as wax solvents, have practically eliminated the use of clearing oils.

Following dehydration in any of the butyl alcohols or dioxan, no clearing reagent is used, because these reagents are solvents of paraffin. They do not render the tissues appreciably transparent.

Dehydration in Solvents of Paraffin

THE BUTYL ALCOHOL METHOD

Normal and tertiary butyl alcohol have been introduced into microtechnique in recent years and show much promise as dehydrating and infiltrating agents. Normal butyl alcohol, also designated butanol, was the first of these higher alcohols to be used extensively. Lang's careful experiments have shown that a miscibility curve of the three components of the dehydrant may be used to ascertain the composition of solutions for an ideal dehydrating series. For critical cytological work, follow Lang's miscibility curves (Lang, 1937) in making up a series. The following series is a simplification that has been found to give excellent results in histological and anatomical work.

Grade number	n-Butyl alcohol (Butanol)	Ethyl alcohol	Water
1	10	20	70
2	15	25	60
3	25	30	45
4	40	30	30
5	55	25	20
6	70	20	10
7	85	15	0
8	100	0	0

Note that each of the first six grades consists of three ingredients. The last two grades are anhydrous. Use new anhydrous butyl alcohol for Nos. 7 and 8. After being used once, No. 8 may be used to make up any of the first six grades.

After an aqueous killing fluid, wash or rinse the tissues in water, dehydrate in alcohol in the usual manner to 30%, then transfer to the above reagent 1 and follow the series. After *FAA* or other fluids having a water content of about 50%, rinse in 2 changes of 50% alcohol and begin the n-butly series with No. 2, in which the water content is 60%. With many histological subjects good results can be obtained by dehydrating to 50% in steps of 10%, then continuing in n-butyl series 3, 5, 7, and 8.

Tertiary butyl alcohol (*TBA*) is regarded by some workers as the most ideal dehydrating reagent of any thus far used (Johansen 1940). Unlike the two other butyl alcohols, its odor is agreeable. The cost is at present much too high for extensive routine work. Tertiary butyl alcohol is used in accordance with the principles of dehydration described in the preceding pages. Dehydrate in ethyl alcohol to 50%, then pass through the following series:

Grade number	95% ethyl alcohol	Absolute ethyl alcohol	TBA	Water
1	50		10	40
2	50		20	30
3	50		35	15
4	50		50	
5		25	75	

Make three changes of anhydrous tertiary butyl alcohol and proceed with infiltration in wax.

An unfavorable factor in the use of tertiary butyl alcohol is that it solidifies at 25.5°C., a temperature that is not uncommonly attained in laboratories and stock rooms. Provisions must be made to keep this reagent fluid for immediate use. The low boiling point of 82.8°C. presents some fire hazard.

The butyl alcohols have greatly extended the range of usefulness of the paraffin method by making it possible to cut materials that are rendered hard and brittle by ethyl or propyl alcohol or acetone.

THE DIOXAN METHOD

Dioxan, diethylene dioxide, is becoming widely accepted as a dehydrating agent and paraffin solvent in the embedding of plant materials. This reagent is miscible with water and may therefore be progressively substituted for water in the tissues. Unlike the vigorous dehydrating action of the alcohols or acetone, the substitution of water by dioxan is not associated with great plasmolyzing stresses. This fact permits dehydration by rapid substitution. Tissues do not become excessively brittle, and the histological details obtainable are equal to those obtained by other methods. The dioxan method requires much fewer separate operations than does any other method, and the operations may be at widely spaced intervals, thus reducing the burdensome routine of handling the specimens many times at frequent intervals.

Kill the material in the desired formula. After the optimum fixing interval, wash in water if required by the formula. Animal tissues are said to be transferable directly from the wash water into pure dioxan, but plant cells are plasmolyzed by such treatment.

Materials that were washed in water are transferred through the following three grades at 4- to 12-hr. intervals. Wide latitude in these intervals is permissible.

(1) $\frac{1}{3}$ dioxan
$\frac{2}{3}$ water
(2) $\frac{2}{3}$ dioxan
$\frac{1}{3}$ water
(3) Anhydrous dioxan. Replace the cork with a perfectly dry one.

Make two more changes of anhydrous dioxan after intervals of 4 to 8 hr. Proceed with progressive infiltration in paraffin as described in the next chapter.

Materials that were killed in *FAA.* or in any fluid that is followed by rinsing in 50% alcohol, are transferred through the following series at 4- to 12-hr. intervals. For small root tips the intervals need not be over one hour.

(1) ½ dioxan
     ½ water
(2) ⅔ dioxan
     ⅓ water
(3) Two changes of anhydrous dioxan as in the previous schedule.

Infiltrate in paraffin as descibed later.

The following five-grade series is recommended for delicate or easily plasmolyzed material: 10% "commercial" dioxan in water, 25%, 50%, 75% dioxan at 1- to 4-hr. intervals. Change corks and make two or three changes of anhydrous dioxan at intervals of 1 to 12 hr., depending on the size of the pieces. The time intervals in this series are not critical. Anhydrous dioxan is a solvent of wax, but the rate of dissolving and infiltration can be increased by the addition of 5 to 10% xylene or chloroform to the last change of dioxan.

If anhydrous dioxan is difficult to obtain, a dioxan-normal butyl alcohol series may be used. The method has been tested extensively and is highly recommended. Dehydrate in the foregoing dioxan series to the commercial grade. Transfer to equal volumes of commercial dioxan and commercial butyl alcohol for 1 to 12 hr. Make two or three changes of anhydrous butyl alcohol and proceed with infiltration. A similar dioxan-tertiary butyl alcohol series also is satisfactory.

Regardless of the dehydrating agent and wax solvent that were used, it is desirable to evacuate any residual air that may remain in the tissues at this point. Place the uncorked specimen bottle into a dry jar, use a safety bottle between this jar and the aspirator, and evacuate until no more bubbles come out of the specimens (Fig. 3.1).

During the experimental period following the introduction of dioxan, unsatisfactory results were reported by many workers. Some lots of dioxan produced severe shrinkage; other purchases yielded acceptable, though variable, results. An inexpensive and satisfactory commercial grade dioxan can now be obtained.

Dehydrating agents can be re-claimed after they have been used and have absorbed some water. Commercial grades that contain a low percentage of water can be made anhydrous. It is rarely profitable to re-claim ethyl alcohol. Pour the used liquid over anhydrous calcium sulphate, known commercially as ""Drierite." After several hours, decant and filter the fluid and it is ready for use. If the fluid has become colored from materials extracted from the tissues, distill at the boiling point of the reagent being distilled. This information can be found in a chemical handbook.

Drierite can be regenerated by air drying and then heating in a furnace at 225 to 250° C. for two hours.

5. Infiltration and Embedding in Paraffin Wax

The paraffin matrix in which tissues are embedded serves to support the tissues against the impact of the knife and to hold the parts in proper relation to each other after the sections have been cut. These functions are best performed if all cavities within the tissues are filled with the matrix and if the matrix adheres firmly to the external and internal surfaces of the material. Infiltration consists of dissolving the paraffin in the solvent containing the tissues, gradually increasing the concentration of paraffin, and decreasing the concentration of solvent. The solvent is eliminated by decantation or evaporation, or both, depending upon the character of the solvent and the process used.

Properties and Preparation of Paraffin

The properties of the embedding paraffin are important factors in the success or failure of sectioning. Desirable properties are as follows:

1. Constant and known melting point and appropriate hardness; the waxes used for most botanical work have melting points between 50 to 55°C., with a tolerance of 2° for a given grade.
2. Smooth, even texture, with a minimum of crystalline or grainy structure.
3. Absence of particles of dirt, included water, and volatile or oily components.

Commercial paraffins from different sources differ widely in properties and suitability for embedding. Purchases made from a given source may vary from time to time — some lots giving satisfactory results, whereas other lots, treated by indentical methods, are unsatisfactory. For these reasons paraffins from the available sources should be tested as to melting point, texture, behavior under the casting methods used, and cutting properties with familiar subjects.

In most parts of the United States, the wax obtained from petroleum is known by the name paraffin, whereas in some areas the term wax is used. The two terms are used indiscriminately in this manual.

Most of the paraffins sold for domestic canning have excellent properties but are too soft for sectioning under ordinary room temperatures or for cutting very thin sections. This inexpensive paraffin is satisfactory for sectioning soft materials such as fruits, if sections over 20 μ in thickness are desired. Paraffin of excellent quality and stated melting point can be purchased from biological supply houses, but at rather high cost. Canning paraffin can be used for preliminary infiltration, and the more expensive hard paraffin used for the final embedding. Canning paraffin requires no preparation; the pieces may be put into the oven tank where melting takes place readily.

Bulk paraffin can be purchased in 10-lb. slabs at low cost from petroleum refining companies. This bulk paraffin usually contains considerable dirt but it can be purified easily. Heat a quantity in a pan until it just begins to smoke, then keep over a small flame for at least ½ hr. Avoid heating the paraffin to the ignition point. Pour the paraffin into a tall metal container, such as a tall coffee can, and permit it to cool in a warm place. This permits particles of dirt to settle. Cool until the surface begins to solidify, then decant into the oven tank. The smoking hot wax can be filtered rapidly through dry filter paper. Use a coarse filter paper and keep the sides of the metal funnel warm with a small bunsen flame.

Each purchase of paraffin should be tested by casting a test block into a mold. The paraffin test block should contain no bubbles, opaque spots, streaks or internal fractures. When the chilled block is broken, the fracture should show a grainless or finely granular surface. The paraffin should slice into thin curled shavings, not into brittle granules. Keep a test block at a temperature of 30 to 35°C. for 24 hr.; bubbles and opaque crystalline spots should not appear.

Cast blocks of good paraffin should remain free from internal defects indefinitely, especially if stored at a constant, low temperature. Occasionally, one encounters old blocks that are almost as clear as glass. Some such waxes have adequately fine grain and may cut very well. In other cases, the impact of the knife causes opaque fracturing of the wax, as when ice is struck with an axe. The wax has obviously crystallized into a coarse texture during storage. When a block of such paraffin is melted slowly in the oven and recast into new wax, the tissues may still be in good condition.

The texture and cutting properties of paraffin can be improved by the addition of rubber and beeswax, and a hard wax, such as ceresin wax. Hance's formula is recommended. Dissolve 20 g. of crude rubber in 100 g. of smoking hot paraffin. Cool and cast into slabs like canning wax. Make up the following mixture:

Paraffin .... 100 g.
Rubber-paraffin mixture .... 4–5 g.
Beeswax .... 1 g.

Ceresin wax may be added to the above, 2 to 5% by weight. Heat the mixture until it just begins to smoke, filter through paper, and cool until it begins to solidify before putting into the oven tank. Tissuemat and Parlax are two commercial embedding waxes that have excellent properties.

Hard waxes and synthetic resins need continued study as hardening agents. For instance, Fisher Scientific Company's "Permount" cover glass resin contains a resin of undisclosed formulation. Dissolve 10 cc. Permount in 200 g. melted Tissuemat in a 60–80°C. oven. When the odor of the solvent, toluene, can no longer be detected, pour into a mold. Melt this very hard wax into 10 to 20 times its weight of Tissuemat. The resulting hardened wax yields thin sections, though not in the range of ultra-thin sectioning.

Apparatus

OVENS FOR INFILTRATION

A well-insulated oven with thermostat-controlled electrical heating is the most reliable type. A removable copper tank makes a suitable container for the supply of melted paraffin. The tank can be equipped with a brass petcock, but petcocks develop leakage. A more satisfactory method is to dip out the paraffin with a spoon as needed. Debris settles to the bottom of the tank, and the clear paraffin is used from the top. An incubator oven with a reliable thermostat is satisfactory for paraffin work, but the temperature in different parts of the oven is not the same and must be determined. In the latter type of oven the supply of melted paraffin may be kept in a container with removable cover and dipped out with a spoon kept hooked in the container. If it is possible to have two ovens for infiltration, use an inexpensive wooden incubator oven set for 35°C. for preliminary infiltration.

DEVICES FOR CASTING BLOCKS

Several methods are in use for casting infiltrated tissues into a mold. The most practical mold is a tray or "boat," made of heavy glazed paper, aluminum foil, or aluminum insulating foil. The method of making boats is shown in Fig. 5.1. Fold along the dotted lines (A) and lap the wide side to lock the narrow side as in B. Masses of loose minute objects that have been processed by centrifuging and decantation can be cast in a pyramid mold (C). Aluminum foil is folded best over a wood form. Insulating foil can be folded as easily as heavy paper without a form.

Soak paper boats in smoking hot canning wax until bubbles cease to come from the paper. Remove the boats from the wax, shake off surplus wax and cool the boats on a paper towel. A supply of boats of various sizes can be prepared in advance and used as needed. Boats

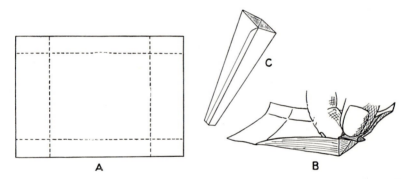

A   B

Fig. 5.1—Method of laying out (*A*) and folding (*B*) paper boats for casting paraffin blocks. *C*, pyramid mold.

of durable, flexible plastic in various sizes will probably be developed. Some brands of plastic ice cube trays have desirable properties.

Some form of hot plate is used to keep the paraffin in the boat melted while the material is being arranged. Electric plates with thermostatic control are available. A sheet-copper table is used in many laboratories. An easily controlled heating table consists of a sheet of ½-in. boiler plate, 6 to 8 in. wide and 18 to 24 in. long, mounted on legs or on a large ring stand (Fig. 5.2).

The warm-pan method must be used for materials that are very small, buoyant, or transparent. The boat is supported on a wire triangle in a pan, which serves as an air bath, heated by a small Bunsen flame (Fig. 5.2).

Infiltration With Paraffin Wax

The following infiltration procedure may be used with any of the common solvents, if the respective specific gravities of the solvent and the wax are taken into account. Paraffin wax, either in solid or melted form, floats on chloroform and also on dioxan. Therefore, either of these solvents, used alone, provides the progressive infiltration that will be emphasized in this chapter. The addition of chloroform or xylene to dioxan, as suggested previously, merely accelerates the dissolving of the wax.

Paraffin sinks in xylene, but if the melted wax is poured into a bottle of cold xylene along the side of the bottle, a solidified layer of wax will remain on top of the solvent during preliminary infiltration at 25 to 35°C. When the solvent is warmed above the melting temperature of the wax, the nearly pure wax will sink to the bottom and envelop the tissues. The abrupt concentration gradient between

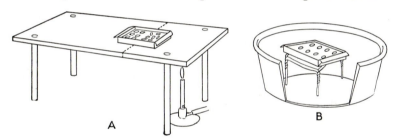

Fig. 5.2—Casting tables: *A,* boiler-plate table (the dotted line indicates the melting zone) ; *B,* warm-pan device.

the wax and the tissues may be destructive to fine detail. The addition of 6 to 10% chloroform by volume to xylene raises the specific gravity sufficiently to float wax chips or melted wax, and a gradual downward diffusion gradient is obtained.

Normal and tertiary butyl alcohol have a lower specific gravity than paraffin wax, and the wax sinks upon the tissues. Flotation of wax can be obtained by the addition of chloroform, 15% by volume to normal butyl alcohol, and 25% to tertiary butyl alcohol, respectively. If chloroform is not added, the addition of wax must be made by small increments, in liquid form, and each increment must be homogenized to prevent the added wax from sinking and enveloping the tissues.

If a mixture of two solvents is used, the bottle should be kept corked until the infiltration is well advanced, to prevent the differential evaporation of the two solvents in the oven.

Pour a teaspoonful of melted paraffin over the cold solvent, where the paraffin solidifies as a layer on top of the solvent. Remove the stopper and place the bottle into the 35°C. oven. The layer of paraffin does not melt, but it gradually dissolves and diffuses downward into the tissues. If all of the paraffin dissolves at this temperature, add more melted wax. If the bottle becomes filled with solution, pour some of the solution into the waste can (*not into the sink!*). Continue the addition of melted wax until a thin layer of undissolved wax remains on top of the solution. The undersurface of the layer of paraffin eventually develops a translucent, crystalline appearance. When this stage is reached, the solvent is obviously *saturated with paraffin at this temperature*. Tissues are not damaged by prolonging this infiltration at 35°C., therefore, this part of the process may be extended over 2 or 3 days.

Transfer the specimen bottle to the 53°C. oven, where the layer of solidified paraffin soon melts and continues to dissolve, and the infiltration initiated at the lower temperature is continued. If the specific gravity of the solvent has not been adjusted as described above, it is best to remove the solid supernatant wax at this point and add wax by small increments at intervals.

If the tissues are not extremely delicate or fragile, whirl the bottle *gently* until the liquid is homogeneous, as shown by the absence of refraction waves within the liquid. At intervals of 1 to 4 hr., pour off one-half of the homogenized solution into the waste can. Replace the decanted liquid with an equal volume of melted soft paraffin, and replace the bottle into the oven quickly. After four or more such partial replacements, pour off all the paraffin-solvent solution, which now consists mostly of paraffin, and replace with pure paraffin. After 1 to 4 hr. make another complete replacement and make a *button* test. Cast a button of paraffin about the size of a silver dollar by pouring some of the paraffin from the tissues into a pan of cold water. Promptly replace the specimen bottle into the oven. Allow the test disk to cool thoroughly. The cooled test button should not be greasy. Chew a piece of this paraffin. The presence of even a slight trace of xylene or other solvent is easily detected by taste. Examine for the defects and qualities described on page 31. If the test piece indicates that all the solvent has been removed, make two changes of hard paraffin or commercial casting compound at 1- to 4-hr. intervals. The material is then ready to be cast into a mold.

The use of two, or even three grades of paraffin with different melting points, used successively during infiltration, has been suggested in the literature. If a laboratory is equipped with three ovens, maintained at 40, 50 and 54°C., there may be some point to the successive use of waxes having those melting points, but if the three waxes are used at the same temperature, there seems to be little basis for the procedure. It can be demonstrated easily that high-melting-point wax that contains a trace of solvent becomes low-melting-point wax. The progressive method outlined in this chapter may be used with only one grade of wax; however, the use of inexpensive canning wax for preliminary infiltration, followed by casting in a high grade filtered wax or compounded formula, is economical and entirely satisfactory for most tasks.

A vacuum oven is an aid in the infiltration of difficult material. Cavities that resist evacuation at low temperature can be exhausted at 50°C., when the material has progressed to approximately equal volumes of wax and solvent. The low boiling point of tertiary butyl alcohol makes the use of a vacuum oven impossible with this solvent until the solvent has been almost completely replaced by wax.

Most material can be adequately evacuated by the time it is in the final solvent, and the dissolving wax will then diffuse into all spaces occupied by the solvent.

CASTING INTO A MOLD

Assume that the infiltrated material is in the final change of pure paraffin. If the oven has cooled because of frequent opening, the paraffin in the specimen bottle may have congealed. Heat the neck and upper portions of the bottle in a Bunsen flame. Never heat the *bottom* of the bottle because the tissues resting on the bottom will be overheated and ruined. Apply only enough heat to liquefy the paraffin. Slight heating repeated at 10-min. intervals is safer than melting at one heating. If you have not yet provided means of identification, write the designation of the given lot of material on a ½-cm. square of paper and put into the bottle. Select a paper or foil boat that will accommodate the pieces in the specimen bottle, without wasting space or wax. Heat one end of the casting plate with a small flame and place the empty boat at the *melting zone*. Pour the paraffin containing the tissues into the boat. Arrange the pieces with a bristle. A warmed needle may be used, especially to move pieces that have become frozen into unsatisfactory positions. Slide the boat over the edge of the melting zone toward the cold end of the plate as fast as a row of pieces is arranged. When the pieces are suitably oriented, move the boat to the cold end of the plate. Sweep the Bunsen flame rapidly over the surface of the paraffin; this permits contraction on the upper surface while the bottom of the paraffin is cooling, thereby preventing the formation of cavities. As soon as the paraffin is hardened enough to keep the pieces of tissue from moving, float the boat in a pan of cold water and brush with the Bunsen flame again. Allow the surface to solidify, and submerge the boat in the water, holding it under with a staining jar cover or other weight. When the paraffin is thoroughly cooled, peel off and discard the paper boat.

Some purchases of paraffin wax, and occasional lots of proprietary compounded waxes, develop white areas or bubbles if kept at room temperature after casting. This can be minimized or prevented by storing the cast blocks in a refrigerator for several days. If the spots appear upon subsequent exposure to room temperature, permanent low temperature storage should be provided.

When using the warm-pan method (Fig. 5.2 B), place a paraffin-soaked boat on the triangle, and warm the bottom of the pan with a Bunsen flame of such size that a layer of paraffin in the boat is kept just melted. Pour the material into the boat, and arrange the pieces, occasionally flaming the top surface of the paraffin. There is little danger of overheating the material; hence the operator need not hurry in arranging the pieces. When the pieces are arranged satis-factorily, remove the burner, and pour cold water into the pan until the water level is slightly above the bottom of the boat. The pieces become hardened into place quickly. Sweep the flame over the surface of the paraffin to permit internal contraction. Complete the hardening as in the former method.

The spacing of pieces in the block depends on the size of the pieces. Root tips and small pieces of leaf can be spaced 5 mm. apart; large pieces of stem require more supporting paraffin during section-ing and should therefore be spaced at least 1 cm. apart. Very slender root tips, coniferous needles, and similar objects may be blocked in groups of three or more pieces laid parallel so that they can be microtomed simultaneously.

RECASTING

Poor paraffin of a cast block can be replaced with good paraffin, or poorly arranged material can be rearranged and recast, and exces-sively large pieces can be trimmed and recast. Cut the pieces out of the block, trim the pieces if desired, cut away excess paraffin if it is of bad quality, and drop the pieces into a bottle of melted paraffin in the oven. Do not apply extra heat; the temperature should not exceed 53 to 54°C. When the old paraffin has amalgamated with the new wax, make at least one change into new casting wax and cast into blocks.

REINFILTRATION

Poorly infiltrated tissues can sometimes be salvaged by reinfiltration. This should not be attempted if there has been excessive collapse of cells, a frequent result of poor infiltration. Cut the pieces out of the paraffin block, trim away excess paraffin, and drop the pieces into anhydrous dioxan, normal butyl or tertiary butyl alcohol, xylene or chloroform. After 24 hr. at 35°C. transfer to the 53 to 54° oven and continue progressive infiltration. A vacuum oven may be used for such salvage operations.

Cast blocks should be stored under conditions that minimize damage to the tissues and to the texture of the paraffin. Trim the edges of the cast block so that both surfaces are flat. Store in a stout manila envelope or small cardboard box bearing adequate identification data. If several blocks are stored in one container, use thin cardboard separators. Box containers should be stacked so that the blocks lie flat. Stout envelopes support the blocks well enough to permit filing the envelopes in the vertical position in a filing cabinet. Storage temperatures should be low enough to prevent bending of paraffin blocks.

6. Microtome Sectioning of Material in Paraffin

Material embedded in paraffin is almost invariably cut with a *rotary* microtome, in which the knife is stationary and the piece of tissue is moved up and down past the cutting edge. Cutting is accomplished by wedge action, like the action of a chisel or a plane. After a section has been cut, and the tissue carrier has passed the knife on the upstroke, an automatic feed mechanism advances the tissue carrier forward, and another section is cut. Successive slices remain attached to each other, forming a *ribbon* of paraffin. The successive sections cut from a piece of tissue are thus kept in *serial* order, and this order can be preserved throughout the processing of the slides. From serial slices of an organ of a plant it is possible to reconstruct the external or internal structure of the organ, of a tissue system, or even of a single cell. As an example of serial sectioning we may use the homely illustration of a loaf of bread, cut into slices and the slices laid out in order.

The piece of material to be sectioned is fastened to a mounting block, which is clamped into the microtome. Inexpensive mounting blocks can be made of hard, porous wood, such as oak or ash. The most useful sizes range from 1 by 1 by 2 cm. to 2 by 2 by 3 cm. Soak the blocks in hot canning wax. Bakelite and other plastics make excellent blocks, preferred to wood because plastic blocks do not compress when clamped into the microtome (Fig. 6.1 *A*). Wood or plastic blocks are satisfactory for most work, being inexpensive and sufficiently rigid for sections over 6 μ in thickness. For routine sampling of material, many pieces of tissue can be mounted on separate blocks, the mounting blocks numbered by means of string tags, and test sections made from each piece. The mounted blocks can be kept with the proper batches of embedded material until staining trials establish which block has the desired stage.

Metal mounting disks (Fig. 6.1 *B*) afford greater rigidity than plastic blocks and are indispensable for cutting very thin sections, or for sectioning large pieces of firm material. Disks remain cold longer after chilling, thereby keeping the paraffin block cold for sectioning. Disks are much more expensive than homemade blocks, and most laboratories have a limited supply, making them unsuitable for the routine sampling method described above. Some microtomes have a built-in tissue-mounting disk on a ball-and-socket joint. This device is satisfactory for work in which each piece of material is used up at one cutting, thus emptying the carrier for the next piece or for the use of other workers. For class use or for work requiring much sampling of diverse materials by several workers, removable disks or blocks are much more desirable.

To fasten a piece of material on a mounting block, trim the paraffin around the material so that the cutting plane is established.

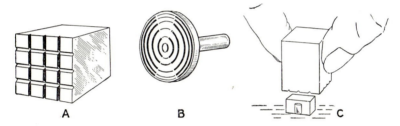

A   **B**   **C**

Fig. 6.1—Mounting of tissues on object blocks: *A*, wood or plastic block with scored surface; *B*, metal object disk; *C*, method of orienting paraffin block and fastening to object block.

Lay the cutting face on a clean surface. Heat the mounting block, press it firmly on the back of the specimen, and hold in contact until the wax cools (Fig. 6.1 *C*). Build up a fillet of paraffin around the specimen to afford firm bracing for the tissues (Fig. 6.2). Cool thoroughly before sectioning.

Some workers of acknowledged skill use a heavy knife for classwork, for routine preparation of teaching material, and for research. Other workers of equal ability use either a knife or a razor blade in accordance with requirements of the work in hand. For sections ranging from 8 to 15 μ in thickness, a sharp razor blade in a suitable, rigid holder will match the work of the heavy knife. The thick type of razor blade (Enders or Christy) can be stropped, used repeatedly, and discarded when stropping no longer restores the edge. In a course in microtechnique, for the routine preparation of teaching material, and for many research tasks, razor blades are economical and entirely satisfactory.

For cutting very thin sections, uncommonly thick or large sections, or for tough materials, the heavy microtome knife is indispensable. The greater rigidity of a knife permits sectioning of material with which a flexible razor would chatter, cutting sections of uneven thickness.

The sharpening of a microtome knife is a laborious process, best learned by observing a demonstration. If a knife is badly nicked or bowed in the center it is best to send it to the manufacturer for grinding. A straight edge and a new correct bevel of the cutting edge are thus established. With a properly ground knife, occasional stropping restores the edge for a considerable time, depending on the hardness of the material being cut. A honing back is a longitudinally split metal cylinder that is slipped over the thick back of the knife for honing or stropping. The diameter of the cylinder determines the angle of the honed wedge. The metal of the cylinder is usually softer than the knife and wears away faster by honing. When the cylinder is appreciably worn, or unevenly worn, a new one should be fitted and a new wedge angle honed on the knife. It may be necessary to hone the knife on a fine gray hone using soap suds as a lubricant. When the fine hone and strop fail to restore the edge, and it is not possible to have the knife machine-ground, a new wedge and cutting edge can be established as follows. Place the cutting edge on the coarser yellow hone with the knife vertical, and make one *light* stroke, removing the cutting edge completely. Then lay the knife flat on the hone and stroke with long oblique strokes, alternating the two sides, until the two sides of the new wedge meet. Examine periodically with a microscope. Hone on the fine gray stone until the edge consists of uniform, minute serrations. Strop as usual before using.

Factors Influencing Sectioning

Successful sectioning in paraffin depends upon a number of interacting factors, the most important of which will be discussed briefly.

QUALITY OF THE PARAFFIN

The hardness of the paraffin should be appropriate for the character of the tissues, the desired thickness of the sections, and for the room temperature at which the cutting is done. The paraffin should have a grainless or very fine-grained texture, should be free from bubbles and opaque spots, and should contain no grit or other debris.

PROPER INFILTRATION

Improperly infiltrated material breaks out of the paraffin block. Examine the cut face of the material with a hand lens or a binocular dissecting microscope. Crumbling within the tissues may indicate inadequate penetration by paraffin during infiltration or may be the result of excessive hardness or brittleness of the tissues. Breaking out of entire sections from the paraffin ribbon indicates poor adhesion between external surfaces of the piece of tissue and the paraffin. Inadequate infiltration may be due to incomplete dehydration or excessively rapid infiltration. The remedy lies in reinfiltration.

ORIENTATION OF THE MOUNTED MATERIAL

The paraffin around the piece of material should be trimmed rectangular, with the material approximately centered laterally in the paraffin. If the tissue is not vertically centered in the paraffin, the thicker layer of paraffin should be at the top, affording support against the pressure of the cutting action (Fig. 6.2 *F*). Trim the upper and lower edges of the wax so that the sections are close enough to each other on the slide for efficient use of the slide and cover glass. See Fig. 6.6 for efficient placement of sections. The edge which approaches the knife should be parallel to the knife (Fig. 6.3 *C*). For most paraffin sectioning the knife is placed at right angles to the vertical motion of the paraffin block. The other angle to be considered is the declination, or the tilt of the flat face of the knife toward the tissue (Fig. 6.3 *A, B*). This angle must be determined by trial.

RIGIDITY OF MOUNTING

The piece of tissue should be firmly attached to a mounting block or disk and supported, especially on the edge away from the knife, by a generous layer of paraffin (Fig. 6.2 *E*). The mounting block, the knife, and the knife carrier must be firmly clamped into place. Inadequate rigidity of the tissue mounting or of the knife results in alternate sections of unequal thickness. This can often be recognized in the ribbon but usually becomes evident during staining. The thicker sections will be more deeply stained than the alternating thin ones. In sections of a large stem there may be alternate deeply stained and lightly stained bands in each section.

TEMPERATURE FACTORS

Cutting is influenced by the temperature of the paraffin block, of the knife, and of the room. If the temperature of one or more of these factors is too high, compression of the sections occurs on impact with the knife. If the temperature is too low, the sections may curl, or successive sections may not adhere and thus fail to form a ribbon. Thick sections are relatively more tolerant to higher working temperatures than are very thin sections. A heavy microtome knife permits a higher temperature than does a razor blade. If the temperature is too high for the grade of paraffin being cut, cool the mounted

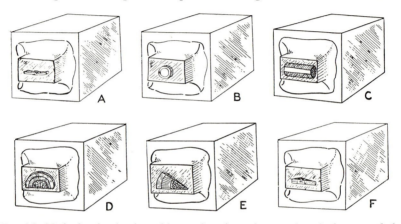

Fig. 6.2—Methods of orienting objects of various shapes: *A*, a leaf mounted for cross sections; *B*, a small stem or other cylindrical organ mounted for cross sections; *C*, for longitudinal sections; *D* and *E*, sectors mounted for cross sections; *F*, sector of large herbaceous stem mounted for radial sections.

paraffin block and the knife or razor-blade holder in a pan of ice water. Align the tissues and knife in the microtome quickly, and cut sections until the paraffin becomes too soft, when the cooling should be repeated. Knife cooling devices are described by Johansen (1940). If a refrigeration room is available, perfect control is possible by setting up the microtome in the cold room and warming a zone around the knife with a desk lamp. The author's department has a cooled microtome room with a floor space of 5 by 7 feet and a height of 8 feet. A small compressor is mounted above the room on a heavy false ceiling. Cooling coils are suspended on the ceiling. Gravity cooling provides a thermostatically controlled constant temperature of 65°F. Two microtomes and two sets of accessories are provided, and two operators have ample space. This room makes the research workers, students in the technique course, and the technicians completely independent of seasonal conditions.

HARDNESS OR BRITTLENESS OF THE MATERIAL

If the above precautions are observed and satisfactory sections and ribbon are not obtainable from dry blocks of tissues, try the warm-water treatment. Plant tissues embedded in wax are not impervious to warm water. If the cutting face of a block of embedded tissues is trimmed to expose the tissues and soaked in water, the paraffin becomes translucent, water penetrates the tissues and renders many hard or brittle subjects soft enough to permit the cutting of excellent sections. Mount a specimen on a metal disk or on a block of plastic, trim as above, put into a beaker of water, and keep in a 35 to 40°C. oven for 12 hr. Objects mounted on wood blocks should be inverted in a vial of water, so that the tissues are submerged. The extent of softening should be tested after 12 hr. by cooling the material to proper cutting temperature and making trial sections. If the tissues are not soft enough, return to the oven for another 12-hr. interval, and test again. If a drop of safranin is added to the water in which the tissues are soaked, the penetration of the dye provides a good index of the depth of water penetration. Some materials, especially improperly infiltrated tissues, crumble and break out of the paraffin with this treatment. After treatment, material cannot be returned to dry storage because the wet tissues become disorganized on drying. If the hot-water method does not yield sections, the material is probably too hard to cut by the paraffin method.

The Operation of the Rotary Microtome

Having studied the foregoing discussion of some general factors that influence paraffin sectioning, we may turn to the specific operations involved. The operation of the microtome can be learned best by observing an experienced worker. Study the diagrams furnished by manufacturers, and examine your particular instrument with a view to understanding the operating principle and interaction of its parts. Some general suggestions are applicable to the operation of most types of instrument. With the tissue carrier at the upper limit of its travel, and the knife removed or at a safe distance from the path of travel of the tissue carrier, clamp the mounting block bearing the tissues into the object clamp. Manipulate the universal joint of the clamp until the forward face of the trimmed paraffin block, or the desired plane of the sections, is parallel to the knife-edge (Fig.

6.3 *C*). Move the knife carrier forward and the tissue carrier downward until the material *almost* touches the knife-edge in its downward travel. Check the setting of the thickness gauge. Make sure that the wedge-like cutting edge is tilted to have proper clearance on the return stroke (Fig. 6.3 *A, B*). This angle must be determined by trial. Inadequate clearance results in compression of the tissues by the forward flat face of the knife or by the edge of the razor-blade

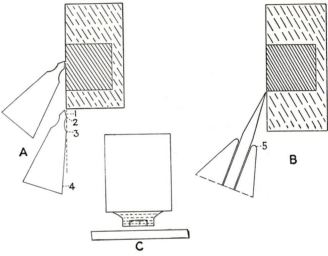

FIG. 6.3—Orientation of tissues in relation to the knife: *A,* knife with the ground cutting wedge and the hollow grind exaggerated to show necessary clearances of angles 2 and 3 and faces 1 and 4; *B,* razor-blade holder, showing declination necessary to clear edge (5) of the clamp; *C,* top view of mounting block, paraffin block, and knife-edge.

holder (Fig. 6.3 *A, B*). Too much angle results in a scraping action rather than a chisel action of the knife-edge.

Having checked the above points, turn the operating wheel slowly, and at the *top* of each upstroke, turn the hand crank of the feed mechanism one revolution, until each downstroke removes a complete slice. Clean the knife-edge by drawing the thumb and forefinger along the front and back faces of the knife, and proceed with the ribboning. Operate the wheel at such speed that there is no marked compression of each slice and successive slices adhere to form a straight ribbon. Note that an experienced worker does not turn the wheel at uniform velocity during a revolution. As the tissue approaches the knife, a snap of the wrist increases velocity considerably at the moment of contact between the tissue and the knife. This speed promotes clean slicing and minimizes compression and wrinkling. At the moment of contact the feed mechanism is disengaged and the only wear is on sliding surfaces. High speed during the entire revolution makes violent impact between the hardened steel pawl and the much softer ratchet wheel, and the pawl may skip a tooth, or strike the top of a tooth. Excessive speed therefore produces excessive wear and is inexcusable, unless the laboratory is lavishly financed. A motor drive, in the hands of the untrained "hired help" that is sometimes used, damages the feed mechanism.

A curved ribbon may be the result of one or more of the following conditions:

1. A dull spot on the knife; shift the knife laterally in its holder or replace with a good knife.

2. The upper and lower edges of the paraffin block are not parallel; trim with a razor blade.

3. The lower edge of the paraffin block is not parallel to the knife-edge; adjust the object clamp.

4. The piece of tissue is not centered laterally in the paraffin; trim the unequal side.

5. The piece of tissue is of irregular shape and bulk. In Fig. 6.4*C* the comparatively empty right side of the paraffin will compress more than the left side, producing a curved ribbon (*B*). This may be corrected by trimming the upper face of the paraffin block (along the dotted line in Fig. 6.4*C*).

The method of straightening a slightly curved ribbon on the slide is described later. Handle the ribbon with a small brush. Do not permit a needle or scalpel to touch the knife-edge. The slightest contact will turn the fine cutting edge. For beginners a quill-shanked brush is the safest implement. Lay the segments of ribbon, in the order of removal from the knife, on clean, lintless black paper, and keep in a cool dust-free place until you are ready to attach them to slides. For handling long ribbons in serial order, cylindrical ribbon holders are manufactured. Their operation is obvious from the catalogue illustrations. The foregoing brief outline of the operation of the rotary microtome should be supplemented by observing the methods of experienced workers. Skill can be acquired only by experience with a wide range of subjects, and a thoughtful analysis of failures.

The condition of the cells and tissues in the ribbon can be judged with considerable accuracy. Examine the ribbon with a magnifier or binocular microscope. The paraffin should be firmly attached to external surfaces and should fill all wrinkles, folds, and visible cavities. Inadequate infiltration may be one cause of separation of the tissues from the paraffin or crumbling within the tissues. If there is abundant ribbon and if seriation need not be maintained, melt a piece of ribbon on a dry, used slide, and examine quickly with a microscope. A magnification of 400 can be used. It is possible to see the chromosomes at metaphase and even at early prophase in onion root-tip cells. The position of chloroplasts in cells can be observed; the degree of granularity of cytoplasm, vacuolation, and plasmolysis can be estimated. The success of the processing can therefore be

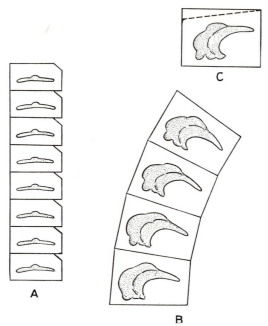

Fig. 6.4—Paraffin ribbon: *A,* straight ribbon, notched if desired by trimming one edge of the paraffin block as in Fig. 15.1*L; B,* curved ribbon; *C,* trimming of paraffin block along dotted line to correct curvature.

judged at this stage in accordance with the criteria discussed in Chap. 11.

The optimum thickness of section for any specific subject can be ascertained at this point. An experienced worker can make a good guess, subject to verification, by examining the ribbon. Study Fig. 6.5 *A,* a perspective sketch of a portion of a leaf with both layers of epidermis omitted. Assume that sections have been cut 20 μ in thickness. Note in *B* that a section of this thickness would encompass two or three layers of narrow columnar palisade cells, and considerable portions of interwoven spongy parenchyma. In such thick sections it is difficult to ascertain the limits of individual cells and the true location of bodies within the cells or mycelium in the tissues. A section 5 μ thick would include adequate longitudinal slices of palisade cells, but the sections cut from various portions of the irregular spongy cells would appear as separated fragments (Fig. 6.5 *C*). Sections of approximately 10 μ might be a good compromise by showing enough of the spongy cells to indicate continuity of contact.

When cutting tissues infected with a filamentous fungus, it is often necessary to make some apparently excessively thick sections in order to include sufficiently long strands of the mycelium. Lily ovules

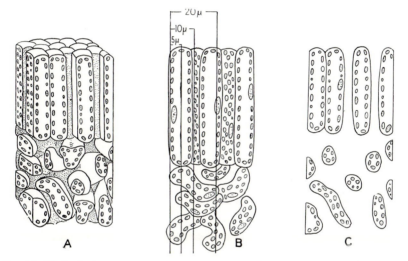

Fig. 6.5—Method of ascertaining the appropriate thickness at which sections should be cut: *A*, perspective view of leaf tissues; *B*, respective numbers of cell layers included in sections of different thickness; *C*, disjointed appearance of leaf tissues in excessively thin sections.

in the four- to eight-nucleate stage must be cut 15 to 20 μ thick to include the complete set of embryo-sac nuclei in a sufficiently high ratio of slides. Onion root-tip sections of 8 to 12 μ include enough chromosomes of the complement to show their approximate number, as well as the organization of the meristematic tissues and their derivatives, but it is desirable to have some slides of 4 to 6 μ to show certain details. Optimum thickness is a compromise between transparency and clearness of separate structures, and the preservation of the relationship and continuity of associated structures.

The very thin sections that are essential for electron microscopy can be made with the attachment supplied by the American Optical Co., Buffalo, N. Y., for their standard microtome, and with special microtomes that are advertised in current journals.

Affixing Paraffin Sections to the Slide

Paraffin sections in the form of a ribbon are fastened to a glass slide with an adhesive prior to staining. Adhesion is influenced by several factors, the most important being the following:

1. Perfectly clean slides.
2. An adhesive (fixative) suitable for the particular material.
3. Proper flattening of the sections by heat.
4. Complete hardening of the adhesive, which makes it insoluble in the reagents used in staining.

New slides should be cleaned, although they may seem to be clean. Use a soapless dish-washing detergent in 70% alcohol. The present favorite in this laboratory is a concentrated liquid detergent, two drops of which in 200 cc. of 70% alcohol makes an effective cleaner. Cleaned slides develop a film on standing, therefore it is best to clean them shortly before using. Used slides can be cleaned with little effort and represent a considerable saving. Slides that have balsam or paraffin on them should be soaked in lead-free gasoline for several hours, wiped dry and cleaned with the detergent. Examine used slides for excessive scratches and surface corrosion. Greasiness of the slide prevents adhesion. To test for greasiness, put two drops of distilled water on the slide and spread with a scalpel. The water should spread out thin on the glass, and not roll inward like water on a hot plate.

The most extensively used adhesives are: gum arabic, obtained in granular form; commercial albumen, such as "Albusol"; fresh egg white; granular or sheet gelatin. Numerous formulas are given by Gray, and elsewhere in the literature. This chapter will present only formulas based on egg albumen and gelatin, by far the most popular and easily available adhesives.

Egg albumen: Drain the white of an egg into a graduated cylinder. Add an equal volume of water and the same volume of glycerin. Add 0.5% sodium benzoate or 1% sodium salicilate. Homogenize in a Waring blender or similar device. Filter and store most of the supply in a refrigerator. Keep a small quantity in the laboratory in a dropper bottle, using a toothpick as an applicator. Place a small drop on the slide, smear into a thin film with the ball of a finger, flood with water, and float the ribbon.

Gelatin: Dissolve 1 g. granular or sheet gelatin in 100 cc. water at 30–35°C. Add 0.5 g. sodium benzoate or 2g. phenol. The addition of 15 cc. glycerin is optional. This stock solution can be used by several methods.

Method 1. Smear a thin film of adhesive on the slide. Flood with 4% formalin and float the ribbon. (Commercial formalin contains 40% formaldehyde gas.)

Method 2. For this one-solution formula, dilute the stock solution with water and add formalin to 4% equivalent. A few trials will determine the dilution that will hold the sections and not leave an excess of stainable gelatin under the sections and on the surrounding glass. The dilution may be as much as 1 vol. stock to 10 vol. water.

Method 3. An old formula that contains both chrome alum and formaldehyde as hardeners has been improved by Weaver by the use of a fungicide and a bactericide, which prevent spoilage of the gelatin.

Solution A	Solution B
Gelatin .... 1 g.	Chrome alum .... 1 g.
Calcium propionate (Mycoban) .... 1 g.	Formalin (40%) .... 10 cc.
Benzalkonium chloride (Roccal) .... 1 cc.	Water .... 90 cc.
Water .... 100 cc.	

Mix 1 vol. Solution A with 9 vol. (or more) Solution B; float the ribbon and proceed as described below.

Decide on the number of pieces to be put on a slide in accordance with the size of the cover glass to be used (Fig. 6.6). Float the desired amount of ribbon on your choice of adhesive. Warm the slide over an alcohol lamp having a wire screen chimney, or on a warming plate, until the paraffin expands, undergoes a change of luster, flattens out, and *approaches* but does not reach the melting point. Keep the ribbon floating while heating to permit expansion. If the paraffin melts, cellular arrangement and cell details are distorted. An insufficiently heated ribbon does not expand or lie flat on the slide and, therefore, does not adhere well. Tough, woody, or elastic subjects are especially difficult to flatten and to attach firmly. If the heated ribbon is curved, straighten while still warm by pulling the concave ends with a pair of needles. Allow the ribbon to cool, then blot with lintless filter paper. Wipe excess adhesive from around the edges of the ribbon to avoid leaving a ring of stainable adhesive.

The slides are now ready to be dried. Some workers prefer to dry slides at a temperature just under the melting point of the wax; however, most tissues are not damaged by drying the slides in the paraffin oven at the melting point of the wax. The adhesive becomes hardened enough in 4 hr. to hold thin sections of soft materials.

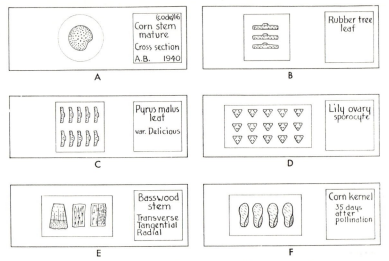

Fig. 6.6—Spacing and arrangement of sections (ribbon) in relation to size and character of subject and size of available cover glass.

Thicker sections of hard material may require 12 hr. The "Technicon" slide drier dries slides in a current of warm, filtered air, and some subjects can be dried and made ready to stain in an hour. Dried slides can be stored, possibly indefinitely, in a dust-free place until you are ready to stain them.

Numbering and Recording of Slides

It is necessary to number or otherwise mark research or demonstration slides after affixing the paraffin ribbon. The serial number or code letter of the material can be scratched on the slide with a diamond, carbide steel or carborundum pencil. Waterproof India ink diluted with an equal volume of gelatin adhesive makes an excellent slide marking ink that will adhere to clean glass through the entire staining process.

In some investigations it is necessary to make a complete series of sections from a piece of material. This may be compared to the many sections from a complete loaf of bread, each slice having a known position in the loaf. A uniform system of placing the sections and numbering the slides should be worked out and rigidly followed. A convenient method is to place the strips of ribbon so that the sections follow the order used in writing, as follows:

1	2	3	4	5
6	7	8	9	10
11	12	13	14	15

The foregoing numbers show the successive order of the sections of a seriation as they are attached to a slide. Another convenient method of designating any given section on the above slide is as follows:

(Row 1)	1	2	3	4	5
(Row 2)	1	2	[3]	4	5
(Row 3)	1	2	3	4	5

The bracketed section is designated as *row 2, section 3*. The numbers 2–3 on the label enable the observer to relocate the section quickly.

A block of tissue may yield so much ribbon that several slides are required to mount the ribbon. In such cases seriation is maintained by mounting the ribbon on the first slide as described above. On the second slide section 16 occupies the upper left hand position as follows:

Slide 1						Slide 2				
1	2	3	4	5		16	17	18	19	20
6	7	8	9	10		21	22	23	24	25
11	12	13	14	15		26	[27]	28	29	30

The number of sections placed on a slide depends on the space available on the slide, the size of the sections, their spacing in the ribbon, and the size of the available cover glasses. Individual sections on any slide are designated by row and section in that row. The bracketed section, which is actually twenty-seventh in the seriation, is identified by the designation *slide 2, row 3, section 2*. Each of the slides in a seriation should be engraved or marked with marking ink, giving the code number of the lot or collection of embedded material, the number of the piece taken out of that lot, and the number of the slide in the seriation. As a specific illustration, assume that from embedded batch 1 of apple leaf you have removed a piece of leaf (piece 1) and sectioned it, and obtained enough ribbon to fill three slides, with sections in serial order. The slides are numbered 1-1-1, 1-1-2, 1-1-3, meaning, in the last instance, batch (lot) 1, piece 1, slide 3. An individual section on slide 3 is then completely identifiable as lot 1, piece 1, slide 3, row 2, section 5. This designation on a drawing or photograph makes it possible to locate the exact section used.

In some subjects the cell size or character of the cellular arrangement makes it impossible to relocate a given cell by row and section. A calibrated mechanical stage should be used to study such material. If the stage revolves, decide upon a reference point on the circular vernier, clamp the stage, and study the slide. Having found a cell which will be sought again for further study, take readings on the longitudinal and transverse verniers, and record in your notes and on drawings. It should then be possible at any future time to put the same slide on the microscope, set the verniers to the recorded readings, and locate the particular cell in the field of view.

7. Staining Paraffin Sections

This chapter is not intended to be a comprehensive treatise on the theory and practice of staining. Historical reviews of the evolution of biological staining and critical discussions of the chemistry of dyes and of staining will be found in Conn's *Biological Stains* (1936). For our purposes it is a safe practical assumption that the staining of cellular structures is based on specific affinity between certain dyes and particular cell structures. This specificity is aided in some processes by a mordant, usually a salt, which enters in some manner into a three-way relationship between the mordant, the dye, and some part of the cell.

This chapter presents a graded series of practical exercises in staining, using designated subjects and a limited number of time-tested stains and combinations. Staining procedures are presented in the form of charts. It is easier to follow a series of operations on a chart than in a written account. Staining processes fall into fundamental types, based on the character of the stains used. Each chart should be regarded as a type chart rather than as a rigid set of specific directions. The sequence of operations should be followed closely, but the time element in some operations should be understood to vary widely. If the *function of each operation* is thoroughly understood, variations of the time element are easily made in accordance with the reactions of the material being stained.

Equipment

Paraffin sections affixed to slides are stained and processed by immersion in reagents in staining jars. Note the various types of jars illustrated in catalogues. The most satisfactory type is the Coplin jar, a vertical jar with grooves that hold the slides in a vertical position. Unlike the horizontal type, Coplin jars occupy little table space, the small opening and the ground-glass lid minimize evapora-

tion, and the slides can be handled more easily in the vertical position. A well-built Coplin jar will hold 9 slides, staggered as shown
in Fig. 7.1 *A*. Some workers place slides into the grooves in pairs,
back to back, but this method does not give the reagents free access
to both surfaces of a slide. For quantity production of slides, screw-
topped jars can be used in conjunction with various types of racks,
holding from 5 to 15 slides (Fig. 7.1 *B*). Staining jars should be
cleaned occasionally in chrome-sulphuric cleaning fluid, washed

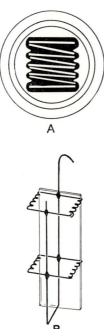

A

B

Fig. 7.1—*A*, Top view of
Coplin staining jar
showing staggered ar-
rangement of nine
slides; *B*, wire holder
for slides.

thoroughly, and rinsed in distilled water. Jars that are to contain
anhydrous reagents must be dried. Assemble a set of at least 15 jars
on a shallow wooden tray, on which they can be carried about or
put out of the way easily. Label each jar in accordance with the
reagents used in Staining Chart I. Do not label water jars, because
the same jar is used for several changes of distilled water and tap
water by pouring out and refilling. If lids are not readily interchangeable, label each lid to correspond with the matching jar. Do not
number your jars. Learn to reason out each step in the staining
process rather than to memorize a numerical sequence of operations.
Each jar should contain enough reagent to cover the slides completely.

Stain Formulas

The existence of several exhaustive formularies makes unnecessary the compilation of an extensive list of stain formulas in this manual. In actual practice only a few of the large number of known stains are used; therefore, the type formulas and specific formulas of only the most frequently used stains are given here. Detailed methods of using these stains are described in the next section.

THE HEMATOXYLIN STAINS

The hematoxylin formulas rank among the most useful biological stains. Hematoxylin is a natural dye, extracted from logwood, *Hematoxylin campechianum* L. The product is purchased in the form of a pale yellow or brownish powder. The certified dye should be specified. Hematoxylin is not a dye in itself, but in the presence of certain alums, which serve as mordants, hematoxylin stains specific cell structures. Numerous formulas and procedures appear in the literature. There are two principal types of formula: (1) the self-mordanting type, in which the hematoxylin dye, the alum mordant, the oxidizing agent, and a preservative are in the same solution; (2) the separate-mordant type, in which the mordant is first applied to the tissues, followed by the application of the dye.

Three of the most useful self-mordanting formulas are given below:

Mayer's Hemalum (Modified)

This stain can be made up conveniently by using stock solutions, thereby requiring only the relatively rough weighing of the alum. Keep on hand a 5% solution of hematoxylin in 95% ethyl alcohol. The dye dissolves slowly at 35°C. and requires about 2 weeks to attain a deep mahogany-red color. Store this solution in a refrigerator, where it seems to keep its properties indefinitely. Make up the stain as follows:

Dissolve 20 g. potassium alum in 1 l. boiling water.
Remove from the heater and add 20 cc. of the above alcoholic hematoxylin, drop by drop.
Add 10 cc. of 2% $NaIO_3$

The stain is ready to use at once. Filter whenever a metallic scum is visible on the surface of the stain in the staining jar. The stain gradually disintegrates and should be made up fresh every 2 to 3 months.

Hematein-Alum (Kornhauser)

Hematein, an oxidation product of hematoxylin, is used in the following formula, which is preferred by some workers to the Mayer formula and its modifications.

0.5 g. hematein.
Grind in a glass mortar with 10 cc. 95% alcohol.
Add to 500 cc. potassium aluminum sulphate, saturated aqueous solution.
The stain is ready to use at once and has good keeping qualities.

Harris' Hematoxylin

1 liter 50% alcohol; 1 g. aluminum chloride; 2 g. hematoxylin crystals.
Heat on a water bath until dissolved.
Add 6 g. mercuric oxide; filter.
Add 1 cc. HCl.

Delafield's Hematoxylin (Slow-Ripening Formula)

100 cc. saturated solution of ammonia alum; add drop by drop 6 cc. absolute alcohol containing 1 g. hematoxylin.
Expose to light in open bottle for 1 week.
Filter, and add 2.5 cc. glycerin and 2.5 cc. methyl alcohol.
Allow to ripen at least 2 months. Filter as needed.

Delafield's Hematoxylin (Rapidly-Ripened Formula, Kohl and James)

Prepare the complete formula as above. Pour into a shallow open dish and expose to a quartz mercury-vapor lamp for 2 hr. Another method of ripening (Neild) consists of exposing the liquid to a Cooper-Hewitt light, for 1 hr., 15 cm. from the light, at 140 volts, 3.3 amp.

The most widely used separate-mordant staining procedure uses the following reagents:

Iron-Hematoxylin (Heidenhain's), (Iron-Alum Hematoxylin)

The mordant consists of a freshly made 4% solution of iron alum (ferric ammonium sulphate). Select clean, transparent, violet-colored crystals, especially avoiding crystals with a rusty coating. Discard the solution when a yellow precipitate develops in the bottle.
The following mordant will keep for months (Lang):

$$
\begin{array}{ll}
4\% \text{ iron alum} \dots\dots\dots\dots\dots & 500 \text{ cc.} \\
\text{Acetic acid (glacial)} \dots\dots\dots & 5 \text{ cc.} \\
10\% \text{ } H_2SO_4 \dots\dots\dots\dots\dots & 6 \text{ cc.}
\end{array}
$$

The stock solution of stain is a 0.5% aqueous solution of hematoxylin. Measure the required volume of distilled water, add a pinch of sodium bicarbonate, about as large as a match head, to a liter of water. Bring the water to the boiling point, remove from the heater and add the dye. Do not boil the solution! Cool promptly and store in a refrigerator. Dilute the stock solution with twice its volume of water for the 4-hr. schedule and with 4 parts of water for a 12-hr. stain. Although the new stain will give satisfactory results, it improves after 2 or 3 days. The stain begins to deteriorate in a few months.

Another type of stock solution consists of a 5 or 10% solution of hematoxylin in absolute ethyl alcohol or 95% alcohol. Dilute to 0.5% in water as needed.

Some satisfactory destaining agents are:

1. Mordant diluted with an equal volume of water.
2. Saturated aqueous solution of picric acid.
3. Equal volumes of mordant and the above picric acid.

Mordanting is sometimes necessary with the synthetic dyes described in the following section. The complex problem of mordanting is well summarized by Popham (1949), and specific suggestions are made throughout the present chapter.

THE COAL-TAR DYES

The coal-tar dyes comprise a large and highly diverse class of synthetic dyes. Their derivation, chemical composition, and properties are discussed in great detail by Conn (1936). Specify dyes that are certified by the Commission on Biological Stains (Conn, 1936). Only the members of this group of stains that are in common use for botanical work will be presented here. Coal-tar dyes are used in a variety of solvents, and the general formulas for the most common stock solutions are as follows:

(1) 0.5 to 1% solution in water, with 5% methyl alcohol optional, as a preservative.
(2) 0.5 to 1% solution in ethyl alcohol, with alcohol concentrations of 50, 70, and 95% and absolute alcohol preferred by various workers.
(3) Saturated solution in clove oil, or
   in equal volumes of clove oil and anhydrous ethyl alcohol, or
   in methyl Cellosolve, or
   equal volumes of clove oil, anhydrous alcohol and methyl Cellosolve.

The following table shows the usual solvents (x) in which the best known coal-tar dyes are used.

Dye	Water	Alcohol, %	Clove oil or Cellosolve
Acid fuchsin (acid)	x	70	
Aniline blue (acid) (= cotton blue)	x	50	
Bismarck brown Y (basic)		70	
Crystal violet (basic)	x	....	x
Eosin Y (acid)		95	
Erythrosin (acid)		95	x
Fast green *FCF* (acid)		95	x
Orange G or gold orange (acid)		100	x
Safranin O (basic)	x	50 to 95	x

The principal botanical uses for the common stains are indicated in the following tabulation:

Cellulose cell walls.
    Hematoxylin (self-mordanting type).
    Fast green *FCF*.
    Aniline blue.
    Bismarck brown Y.
    Acid fuchsin.
Lignified cell walls.
    Safranin.
    Crystal violet.
Cutinized cell walls.
    Safranin.
    Crystal violet.
    Erythrosin.
Middle lamella.
    Iron hematoxylin.
    Ruthenium red (material cut fresh).
Chromosomes.
    Iron hematoxylin.
    Safranin.
    Crystal violet.
    Carmine (for acetocarmine smears).
    Orcein
Mitochondria.
    Iron hematoxylin.
Achromatic figure.
    Crystal violet.
    Fast green *FCF*.
Filamentous fungi in host tissues.
    Iron hematoxylin.
    Safranin O.
    Fast green *FCF*.
Cytoplasm.
    Eosin Y.
    Erythrosin B.
    Fast green *FCF*.
    Orange G or gold orange.

The above tabulations indicate some relationship between the acid or basic character of a stain and its specificity. A basic stain is one in which the *color bearer* is a basic radical; in an acid stain the color bearer is an acid radical. As a rule basic stains are selective for nuclear structures and, in some processes, for lignified cell wall. Acid stains usually are selective for components of the cytoplasm and for unlignified cell wall.

The common clearing oils (clove oil, cedar oil, bergamot oil, and wintergreen oil) usually are used in concentrated form as purchased or thinned with xylene. An inexpensive clearing agent, known as *carbol-xylene* or *phenol-xylene,* consists of 1 volume of melted c.p. phenol (carbolic acid) and 4 volumes of xylene. Phenol should not contain H_3PO_2, which destroys hematoxylin.

Staining Processes

To meet the needs of teachers and beginners, staining processes are arranged in a graded sequence, beginning with the simplest processes, in which the variables and possibilities for errors are reduced to a minimum. The simplest type of stain is a *progressive* stain, in which the intensity of the color imparted to the tissues is proportional to the length of immersion in the stain. Some of the most useful stains of this type have hematoxylin as the active ingredient. In this category of self-mordanting stains are Delafield's hematoxylin, Harris' hematoxylin, and Mayer's hemalum. Many modifications may be found in the literature. The term "hemalum" is used in this manual to refer to any of the self-mordanting alum hematoxylins. The choice among these stains is a matter of personal preference.

HEMALUM (PROGRESSIVE)

The modification of Mayer's hemalum, on which staining Chart I is based, is selective for cellulose, pectin, fungus mycelium in many cases, weakly selective for chloroplasts, strongly selective for metabolic (resting) nuclei, and moderately selective for chromosomes in some subjects. Hemalum may be used without any other stain for meristematic organs, for anther and ovary slides in which a critical chromosome stain is not necessary, and for subjects having but little strongly lignified or differentiated tissues. This stain develops a "metallic" scum on standing. The particles of this scum adhere to the adhesive and to the sections on the slide, therefore the stain should be filtered whenever this scum is evident.

The preliminary processing of slides, prior to immersion in stain, is essentially the same regardless of the stain used. This prestaining process will now be outlined and the procedure is understood to apply when an aqueous stain is used. After the affixed sections have been dried in the 53°C. oven, the sections and adjacent parts of the slide are found to be coated with melted paraffin from the ribbon. Obviously, the first operation is to dissolve this paraffin by immersing the slide in a jar of xylene. If slides are taken directly from the oven, the paraffin dissolves in 1 or 2 min. With cold slides it is better to allow 5 min. The slide is now in a very dilute solution of paraffin in xylene, which is removed by immersing the slide in anhydrous alcohol. As outlined in Staining Chart I, progressive transfer to water is then made through the indicated grades of ethyl alcohol. Isopropyl alcohol or acetone also may be used in most of the staining charts in this chapter. The slide has been *run down* to water, and is now ready to be stained in an aqueous dye. Transfers should be made quickly so that the slides do not become dry. The intervals can be shortened to 30 sec. by moving the slides up and down in the solution with forceps.

The series of reagents in which slides are deparaffined should be replaced when the 30% alcohol becomes cloudy or when the fluid drains from the slides as if the glass were oily, indicating that paraffin and xylene have been carried down the series until the 70% and 30% cannot hold the xylene-paraffin contaminant in solution. The addition of 10% n-butyl alcohol to the anhydrous and 95% grades prolongs the useful life of the series.

The correct staining interval for a given subject must be determined by trial. An experienced worker can make a good guess for a trial slide and make corrections for subsequent slides. One collection of lily ovary killed in Bouin's solution required only 10 min. for a brilliant stain, whereas another collection, fixed in Craf required 1 hr. A collection of lily anther in the microspore stage yielded excellent slides with a 30-min. stain. To determine the correct interval, stain three slides of a subject for three intervals, *i.e.,* 10, 20, and 40 min., respectively. Mark the slides before staining. The sample slides may be held in distilled water and put into the stain at intervals, or they may be put into the stain simultaneously and removed after the desired intervals. After staining, rinse the slides in two or more changes of distilled water, then rinse in three changes of tap water, or in running tap water for 2–5 min. Note that the color in the tissues changes from purple to blue after the transfer into tap water. Hematoxylin gives a reddish-purple color when acid and a blue color when alkaline. The latter color is preferred for the subjects recommended for this first exercise. If the city water in your community does not produce the bluish tinge in tissues that have been stained in hemalum, use 0.1% sodium carbonate for the last rinse. This process may be called *alkalizing.*

At this stage, examine the three test slides that were stained as suggested above. Use a smear microscope, preferably one that has only 10x (16 mm.) and 20x (8mm.) objectives and no condenser. The magnification is adequate and the objectives have such long working distance that they are not likely to be dipped into reagents on the slide. The tissues must not become dry during this quick examination. Nuclei should be blue-black. Cellulose cell walls should be black, whereas lignified cell walls should be nearly colorless. Plastids may be pale blue to blue-black, and cytoplasm blue-gray. If the foregoing structures do not have a deep enough color, transfer the slide from the water to hemalum and give it another interval in the stain, usually as long as the first immersion. Rinse and wash in tap water, examine again and if satisfactory, proceed with dehydration as in Chart I.

If a slide is left in hemalum longer than the optimum period, the contents of the cell may become black, and the details of the wall and protoplast may be obscured. The slide can be destained by brief immersion in dilute acid. The preferred destaining agents are 1 to 5% acetic acid, 0.5% hydrochloric acid, or a saturated aqueous solution of picric acid. Try one minute in acid, wash, alkalize and re-examine with a microscope. When the stain is satisfactory, proceed with dehydration according to Chart I.

Staining Chart I now calls for progressive dehydration of the tissues and the surface of the slide, followed by "clearing." Consult the reference manuals for the various clearing agents in common use. An inexpensive agent is carbol-xylene, the formula of which is given on page 61. Both ingredients must be of high purity. Phenol has a great affinity for water and removes the last traces of water from the preparation. Xylene has nearly the same index of refraction as glass, thus rendering the tissues transparent. High-grade phenol and xylene should not affect the stain even after several days of immersion. Equal volumes of xylene and cedar oil may replace the carbol-xylene.

The final operation consists of cementing a cover glass on the preparation. Have ready a supply of newly cleaned and dried cover glasses. Use a cover glass of generous, but not wasteful, size, with shape and dimensions in keeping with the material to be covered (Fig. 6.6). Discoloration of resin and fading of stain with age proceed from the edges of the cover inward. Have a margin of at least 5 mm. between the sections and the edge of the cover glass. For mounting one section on a slide, or a few sections in a single row, use a ½-, ¾-,

STAINING CHART I *

Progressive Hemalum

Xylene
2–5 min.
(de-waxing)
↓
absolute
(anhydrous)
alcohol
2–5 min.
↓
95%
alcohol
2–5 min.
↓
70%
alcohol
2–5 min.
↓
50%
alcohol
2–5 min.
↓
30%
alcohol
2–5 min.
↓
distilled
water
1–2 min.
↓
Hemalum
5–30 min.
↓
distilled
water
1 min.
↓
Tap water
(see page 62) ⟶

resin and
cover glass
↑
xylene III
5 min.
↑
xylene II
5 min.
↑
xylene I
5 min.
↑
carbol-
xylene
5–10 min.
↑
absolute
alcohol II
5–10 min.
↑
absolute
alcohol I
5–10 min.
↑
95%
alcohol
5–10 min.
↑
70%
alcohol
5–10 min.
↑
50%
alcohol
5–10 min.
↑
30%
alcohol
2–5 min.

---

* The beginner is advised to copy each staining chart on a large card. By means of colored arrows, indicate the sequence of operations used to correct overstaining or understaining.

or ⅞-in. cover glass. For large longitudinal sections of rectangular outline, or for covering several rows of sections on a slide, use a square or long cover glass of such size that there is a margin of at least 5 mm. Caliper all cover glasses, using only those that fall within 0.15 to 0.20 mm. in thickness.

Canada balsam has been the most widely used mounting medium

for many years. Stained sections mounted in balsam may remain in perfect condition for 25 years. However, it is much more likely that the stain will fade, the balsam will become dark yellow, and may even become cracked and opaque like dried varnish. In recent years, numerous synthetic resins have been tried as mounting media. (Lillie, Winkle and Zirkle, 1950). Further experimentation can be expected in the future and the many possible polymers will be tested. The reader is advised to consult the catalogues of biological supply dealers for the currently recommended resins.

The affixing of cover glasses should be accomplished quickly and neatly. Remove a slide from the last xylene, and place with tissue upward on a sheet of dry blotting paper. Working rapidly to avoid drying of the tissues, wipe excess xylene from around the sections, put a drop of resin on the tissues and lower a cover glass obliquely onto the resin. A black background aids in seeing and expelling bubbles. If the size of the drop of resin is correctly gauged, there should be no excess resin squeezed out around the edges or over the cover glass. Newly covered preparations must be used with care because the cover glass is easy to dislodge and the tissues may be damaged. Drying new slides in the 53°C. oven for one or more days hardens the resin somewhat and permits safer handling of the slides.

This is a convenient point at which to discuss the repair of damaged slides. It is possible to salvage a slide that has some sound sections as well as some sections that have been damaged by misuse. Place the slide upside down under a low-power objective and locate the damaged sections. Place a mark over each broken section with India ink. Allow the ink to dry thoroughly, and drop the slide into a jar of xylene. After the cover glass has slid off, rub off the damaged section with a matchstick, rinse in xylene, and mount a new cover glass.

Destaining and Restaining

Slides may be examined for color at several stages in the staining process, in fact from any reagent that is not so highly volatile that the preparation becomes dry during a brief examination. See page 63 for the procedures used to increase or decrease the intensity of the color imparted by hemalum. If the slide is examined out of xylene or carbol-xylene and the stain intensity needs to be increased or decreased, transfer the slide backwards through the dehydrating series to water, and proceed with corrective measures.

It may be necessary to modify the stain intensity of a finished slide that has had a cover glass affixed with balsam or synthetic resin. The cover can be loosened and allowed to slide off by immersing in a jar of xylene as long as necessary. After the cover has slid off, transfer the slide to the absolute alcohol after the de-waxing xylene, run down to water and proceed with the restaining or destaining process.

HEMALUM (REGRESSIVE)

The destaining action of acids on hematoxylin is selective,— cytoplasm is destained more rapidly than cell walls, plastids, and nuclear structures. This fact makes it possible to use a self-mordanting hematoxylin as a stain that is adequately differential for many subjects. The slides are purposely overstained in Delafield's, Harris', or Mayer's hemalum, then destained in acid until the proper contrast is obtained.

The foregoing single stain, using a self-mordanting hematoxylin formula, either as a progressive or regressive stain, deserves more extensive use for routine diagnostic examination of research material. An enormous amount of time and energy can be spent in applying elaborate multiple stains to large numbers of slides, many of which are discarded after a moment's examination. In such a series of slides, stained with a single stain, the few slides having the desired stage can be easily restained if a more diagnostic differential stain is needed.

HEMALUM WITH A "GENERAL" COUNTERSTAIN

The foregoing hemalum stain can be supplemented by a *counterstain,* a stain that has little specific selectivity, but furnishes optical contrast for the principal stain. A counterstain is introduced into the staining series at a place having approximately the same water concentration as the solvent of the counterstain. One of the most useful counterstains is erythrosin. The stock solution contains ½% stain dissolved in 95% alcohol. Referring to Staining Chart II, note that the slide, previously stained to the correct intensity in hemalum, rinsed and alkalized, is put into erythrosin after 95% alcohol. The interval in erythrosin must be determined by trial and may range from a few seconds to 1 hr. This counterstain is removed from different types of material in variable degree by the subsequent dehydration. The final intensity of the pink counterstain depends on the tenacity with which the tissues retain the stain. If the pink color is too dark, it will obscure some of the details stained blue by the hematoxylin. Excess counterstain can be removed by running the slide back to 50% alcohol. More pink can be added as shown on

Staining Chart II. The same slide can be repeatedly destained or restained in the counterstain until exactly the desired effect is obtained. The hematoxylin is not affected during this manipulation.

STAINING CHART II

Hemalum With "General" Counterstain
Pre-Staining Operations and Intervals as in Chart I

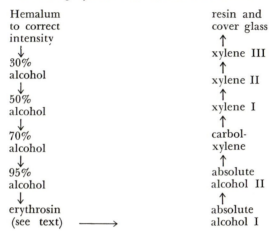

Hemalum	resin and
to correct	cover glass
intensity	↑
↓	xylene III
30%	↑
alcohol	xylene II
↓	↑
50%	xylene I
alcohol	↑
↓	carbol-
70%	xylene
alcohol	↑
↓	absolute
95%	alcohol II
alcohol	↑
↓	absolute
erythrosin	alcohol I
(see text) ⟶	

Other common counterstains used with the above hematoxylins are orange G, gold orange, eosin, fast green, and light green. The underlying principle for applying other counterstains is the same as for erythrosin. Counterstains may also be dissolved in clove oil and applied after the last dehydrating step, omitting carbol-xylene because clove oil is an excellent clearing agent. Counterstains may also be dissolved in water, 50% to absolute alcohol, or Cellosolve and introduced into the series at the corresponding point of dehydration.

HEMALUM AND SAFRANIN

After acceptable results have been obtained with the foregoing single stain and the double stain, undertake the mastery of a double stain having two selective components. One component of the next double stain to be discussed is a self-mordanting hematoxylin; the second component is safranin, which is highly selective for chromosomes, lignin, cutin, and in some cases for hemicellulose. An important feature of this combination is that the hemalum is applied to the desired intensity and remains fixed throughout subsequent processing, whereas the safranin is applied until the material is strongly overstained and then differentially destained.

Staining Chart III begins with a slide that has been stained in hemalum as shown in Chart I; the slide is then immersed in safranin. The interval in safranin ranges from a few minutes to 12 hr. Some collections of young corn stem require at least 1 hr. in safranin. Wood sections cut in celloidin may take up enough safranin in 5 min. to make destaining difficult. Untested material should be tried at intervals of 10, 30, and 60 min. and 8 to 12 hr. After removal from safranin and rinsing in water, all cells of the section are found to be stained deep red, the blue color of the hemalum being masked. Dehydration and differential destaining are accomplished simultaneously by passage through the alcohol series. Safranin is removed from

STAINING CHART III

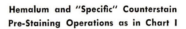

Hemalum and "Specific" Counterstain
Pre-Staining Operations as in Chart I

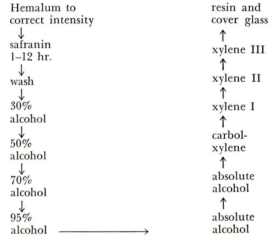

Hemalum to correct intensity

↓

safranin
1–12 hr.

↓

wash

↓

30% alcohol

↓

50% alcohol

↓

70% alcohol

↓

95% alcohol ⟶

resin and cover glass

↑

xylene III

↑

xylene II

↑

xylene I

↑

carbol-xylene

↑

absolute alcohol

↑

absolute alcohol

cytoplasm and unlignified tissues by 50 and 70% alcohol and at a slower rate by similar grades of acetone. Higher concentrations of alcohol and anhydrous alcohol also dissolve the safranin, but 90% acetone and anhydrous acetone have slight destaining action. Acetone, therefore, permits easier control of destaining than does alcohol. Lignified tissues, cutin, and plastids retain safranin throughout suitably rapid dehydration. The correct stain has been attained when lignified cell walls are a clear, transparent red and unlignified walls are blue, with little or no reddish tinge. Chloroplasts may be blue, violet, or red. In order to make chloroplasts red enough to show up clearly, it may be necessary to compromise by leaving too much red in the cellulose walls. If a finished preparation is found to be unsatisfactory, the cover glass can be removed, and the material destained or restained. However, alterations in the intensity of the safranin can be made best after the slide has been examined from carbol-xylene. Carbol-xylene has a very slow destaining action on safranin. Preparations left in carbol-xylene for 4 to 12 hr. show highly critical differentiation of structures having varying degrees of lignification, such as the stratifications in the walls of xylem cells and sclerenchyma.

SAFRANIN-FAST GREEN

The next type of stain combination to be considered has two components, both of which are subject to differential destaining and which react upon each other during dehydration. This staining process is obviously more difficult to control than the preceding processes. As shown in Staining Chart IV, the first stain to be applied is aqueous safranin, in which the preparation is strongly overstained. One hour in safranin is occasionally enough; some woody materials stain well in 5 min. Your previous experience with the hemalum-safranin combination will indicate the safranin-holding capacity of tested materials. The safranin begins to dissolve out during passage through 30, 50, and 95% alcohol. The counterstain, fast green *FCF* in 95% alcohol, is now applied. Both the green stain and its solvent have a differential solvent action on the safranin, and remove it from the unlignified tissues more rapidly than from the lignin, cutin, and chromatin. The interval in green is usually a matter of seconds, rarely as much as 2 min. Correct contrast has been attained when lignin, chromatin, and in some cases cutin are brilliant red, chloroplasts pink to red, and cellulose walls and cytoplasm are green.

The two stains of this combination can be manipulated until the desired contrast and intensities are obtained. If alcohol is used in the dehydrating series, the slide may be placed on a microscope, kept wet with 50% alcohol, and observed until only the lignified elements remain red. The slide is then rapidly carried through the subsequent processes. Acetone is too volatile to permit such examination. With some experience it is possible to judge when the safranin has been destained sufficiently to add the green counterstain. If the stock solution of fast green acts too rapidly for a given subject, the green color will mask or remove the red, and all cells may become stained deep green. In such cases dilute the green stain with 1 to 5 volumes of 95% ethyl alcohol. The slide may be examined best out of carbol-xylene. If red color is still evident in cellulose walls and cytoplasm, carry the slide backward through the series to fast green, double the previous interval in fast green, run upward again to carbol-xylene and examine. This process can be repeated until the desired color contrast between chromatin, lignified walls, cellulose and cytoplasm is obtained.

If the red color is too pale when the slide is examined out of carbol-xylene, transfer to the de-waxing xylene and proceed as with a new slide. The green will be removed in the down series. If the green is too intense at the carbol-xylene stage, back downward to 70% alcohol, in which the green is removed rapidly. Try 10 seconds and carry up to carbol-xylene again and examine.

Several stains can be substituted for fast green in Chart IV. The most commonly used other green stains are light green and malachite green. Several excellent blue counterstains are cotton blue, methylene blue, gentian violet (crystal violet), and aniline blue. Any of these green or blue counterstains can be used in solution in 95% alcohol, in the sequence shown in the chart, or they may be dissolved in 50% alcohol or in clove oil and introduced at the appropriate place in the series. The above safranin-green or safranin-blue combinations

STAINING CHART IV

Safranin-Fast Green
Pre-Staining Operations and Intervals as in Chart I

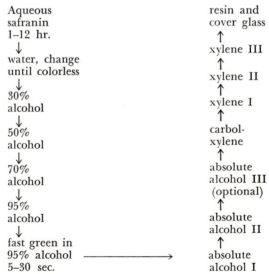

Aqueous
safranin
1–12 hr.
↓
water, change
until colorless
↓
30%
alcohol
↓
50%
alcohol
↓
70%
alcohol
↓
95%
alcohol
↓
fast green in
95% alcohol ――――――→
5–30 sec.

resin and
cover glass
↑
xylene III
↑
xylene II
↑
xylene I
↑
carbol-
xylene
↑
absolute
alcohol III
(optional)
↑
absolute
alcohol II
↑
absolute
alcohol I

serve as excellent cytological stains for many subjects, primarily for the preparation of classroom materials.

THE TRIPLE STAIN (FLEMMING)

The triple stain is of considerable historical interest and is still in high favor in some laboratories. The three components are safranin, crystal (gentian) violet, and orange G (or gold orange). Safranin is intended to stain chromatin, lignin, cutin, and in some cases chloroplasts. Gentian violet should stain spindle fibers, nucleoli during some phases, and cellulose walls. The orange dye acts as a differentiating agent, serves as a general background stain, and stains cytoplasm and in some subjects cellulose walls. All three components are highly soluble in the reagents used in the staining process and are subject to changes of intensity and mutual interaction during most of the process. The correct balance of relative intensities is, therefore, very difficult to control. The process yields spectacularly beautiful slides from the hands of an expert. However, an attractive or gaudy polychrome effect is not adequate justification for the use of an elaborate and time-consuming process. The real test of the desirability of a multiple stain is the specific selectivity of its color components for definite morphological or chemical entities in the cell.

The sphere of usefulness of the triple stain may be judged by a consideration of the stains used in modern cytological research. It is noteworthy that the most critical modern work on chromosome structure and behavior has been done with the iron-hematoxylin stain, with the gentian violet-iodine stain, and with acetocarmine smears. The most reliable work on the spindle-fiber mechanism and spindle-fiber attachment also has been done with the first two stains. As an illustration in the field of anatomy, it will be obvious that in studies of vascular tissues a stain is required primarily to show a xylem-phloem contrast, distinguishing between lignified and unlignified cell walls. This usually is done adequately with a two-stain combination. There is no special virtue in having a delicate orange background for a study of the organization of a vascular bundle or in a section of pine lumber. However, in many cytological problems involving the entire cell rather than merely the actively dividing chromosomes, the triple stain is an indispensable tool. Another legitimate sphere is in pathological studies in which it is desirable to produce polychrome contrasts between a parasite and its host. The object of the above discussion is to emphasize again the view that any elaboration that does not serve a definite, useful function is a waste of time. The triple stain should be kept in its proper place among the diverse tools of the technician.

The three stains used in the conventional process are the following standard stock solutions:

Safranin O, aqueous, or in 50% alcohol.
Crystal violet, or gentian violet, 0.5 or 1.0% in water.
Orange G, or gold orange, saturated solution in clove oil.

Mordanting is necessary for some subjects. After killing in fluids that contain osmic and chromic acids, mordanting is usually not necessary. For materials that do not retain the stains, mordant for 1 to 12 hr. in 1% aqueous chromic acid or in an aqueous solution containing 2% chromic acid and 0.5% osmic acid.

Staining Chart V is intended primarily to show the sequence of operations in a typical schedule. The variability of the time element

STAINING CHART V

Triple   Stain
Pre-Staining Operations and Intervals as in Chart I

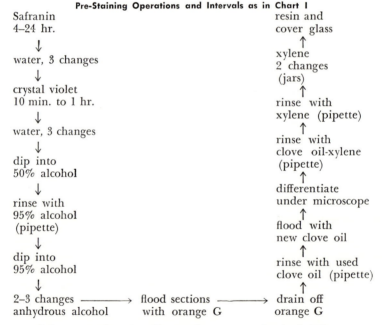

Safranin
4–24 hr.
↓
water, 3 changes
↓
crystal violet
10 min. to 1 hr.
↓
water, 3 changes
↓
dip into
50% alcohol
↓
rinse with
95% alcohol
(pipette)
↓
dip into
95% alcohol
↓
2–3 changes ──────→ flood sections ──────→ drain off
anhydrous alcohol   with orange G   orange G

resin and
cover glass
↑
xylene
2 changes
(jars)
↑
rinse with
xylene (pipette)
↑
rinse with
clove oil-xylene
(pipette)
↑
differentiate
under microscope
↑
flood with
new clove oil
↑
rinse with used
clove oil (pipette)
↑

in any of the operations hardly can be overemphasized. The suggested intervals merely furnish a starting point for experimentation to determine the optimum time schedule for any specific subject.

Many modifications of schedule have appeared in the literature.

Variations in the composition and purity of the component stains have necessitated revisions of schedule to suit the currently available stains. Some workers prefer to differentiate the safranin with very dilute HCl or with acidified alcohol before adding the violet. The acid must subsequently be thoroughly washed out with water.

The gentian violet may be almost fully differentiated in clove oil containing no orange G. The orange can then be added progressively and the violet brought to final differentiation in xylene containing 1/10 to ¼ (by volume) clove oil saturated with orange G or gold orange.

Quadruple-stain combinations using four coal-tar dyes are available in some excellent commercial slides. Conant (Triarch) uses a combination of safranin, crystal violet, fast green, and gold orange; Johansen (California Botanical Materials Co.) uses safranin, methyl violet 2B, fast green, and orange G. These complex processes yield striking preparations but are probably unnecessarily elaborate for most tasks. The advanced worker can obtain details of procedure from the excellent service leaflets of the above manufacturers and from the Johansen manual (1940).

Staining processes using coal-tar dyes are entering a most interesting and important phase of development. Many new organic solvents are being produced by synthetic methods. Solvents that have been little more than chemists' curiosities are now being produced in large quantities and are available at reasonable cost. Some illustrations are the higher alcohols, such as the butyl, propyl, and amyl series, ethyl and methyl Cellosolve, trichloroethylene, and many other solvents. The stains themselves are undergoing constant study and improvement. The possibilities of systematic study, or just plain dabbling, should gratify the heart of the most inveterate experimenter.

IRON HEMATOXYLIN

The next stain to be considered is known as Heidenhain's, or iron-alum hematoxylin. The history of this stain and the names of several investigators who have contributed to its development may be found in the literature. This stain is primarily a cytological stain, used especially for chromosome studies, but it is useful for studies on the cell wall, plastids, and in some studies in pathological histology. The formulas of the mordant (iron alum), the stain (hematoxylin), and the destaining agent are given in the stain formulary. The schedule advocated here and outlined in Staining Chart VI is known as the short schedule, or 4–4 schedule; *i.e.,* 4 hr. in mordant, thorough but quick rinsing in five changes of distilled water at 1-min. intervals, then 4 hr. in stain. The material becomes stained solid black and must be differentially destained. The destaining solution removes the stain rapidly from cytoplasm, less rapidly from plastids, and slowly from chromatin and from the active tips of mycelium. The destaining action should be observed under a microscope and stopped by

STAINING CHART VI

Iron   Hematoxylin
Pre-Staining Operations as in Chart I

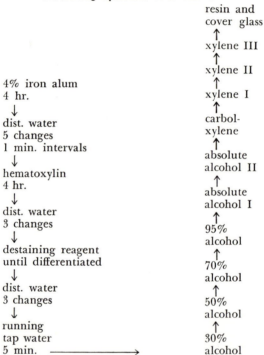

4% iron alum
4 hr.
↓
dist. water
5 changes
1 min. intervals
↓
hematoxylin
4 hr.
↓
dist. water
3 changes
↓
destaining reagent
until differentiated
↓
dist. water
3 changes
↓
running
tap water
5 min. ————————→

resin and
cover glass
↑
xylene III
↑
xylene II
↑
xylene I
↑
carbol-
xylene
↑
absolute
alcohol II
↑
absolute
alcohol I
↑
95%
alcohol
↑
70%
alcohol
↑
50%
alcohol
↑
30%
alcohol

washing first in distilled water, then by prolonged washing in running tap water. Dehydration and subsequent processing follow as shown on the chart. In the finished slide, chromosomes, the chromatin of resting nuclei, middle lamella, and active mycelia should be blue-black, whereas the cell walls and cytoplasm should be practically colorless. A finished, covered preparation may be destained by soaking off the cover glass, running the slide back through the series into water, then immersing in the destaining solution. If a preparation is found to be destained too much, it can be run back into water and mordanted and stained again. As in some other staining combinations, restained preparations are seldom as clear in detail as slides stained correctly the first time.

THE CRYSTAL VIOLET-IODINE STAIN

The next stain to be considered is another cytological stain. It is included here because it yields preparations that are valuable in the teaching of some topics. In teaching mitosis at the elementary college

STAINING CHART VII

Crystal (Gentian) Violet-Iodine
Pre-Staining Operations as in Chart I

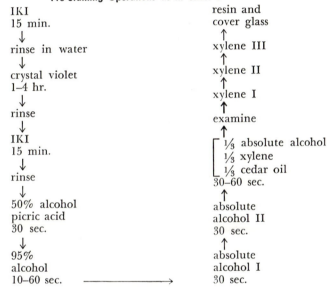

IKI
15 min.
↓
rinse in water
↓
crystal violet
1–4 hr.
↓
rinse
↓
IKI
15 min.
↓
rinse
↓
50% alcohol
picric acid
30 sec.
↓
95%
alcohol
10–60 sec. ⟶

resin and
cover glass
↑
xylene III
↑
xylene II
↑
xylene I
↑
examine
↑
⎡ ⅓ absolute alcohol
⎢ ⅓ xylene
⎣ ⅓ cedar oil
30–60 sec.
↑
absolute
alcohol II
30 sec.
↑
absolute
alcohol I
30 sec.

level, most teachers prefer longitudinal and transverse paraffin sections, stained to show cell walls and tissue organization as well as the more prominent features of nuclear division. Crystal violet stains the chromosomes a brilliant blue-black, on a colorless, almost invisible, cellular background. Such preparations are a valuable supplement to the standard classroom slide.

The procedure given in staining Chart VII is but one of the numerous variants of the process. Probably no other staining procedure is as responsive to the individual touch of the technician, and a specific procedure should be regarded as a point of departure for developing a personal routine.

The following version recognizes that it is cheaper to rinse out stains and mordants in water than in alcohol, and also prolongs the usefulness of the dehydrating series. The stain is a ¼ to ½% aqueous solution of crystal (gentian) violet. The mordant is aqueous IKI (Chap. 9). The picric acid solution is a saturated solution in 50% alcohol. De-waxing and running down to water are carried out in Coplin jars. However, the slides must be carried one by one through the dehydrating series, which must not contain other stains as contaminants. Therefore, a special dehydrating set should be maintained for the violet process, preferably in wide-mouth screw-capped jars.

Differentiation is accomplished in the alcohol dehydrating series. The slide should be agitated in the fluids with forceps. Observe closely in the 95% alcohol, and when visible color no longer comes out of the sections, move rapidly through anhydrous alcohol. Differentiation should be complete when the alcohol-xylene-cedar oil clearing solution is reached. The function of the oil is to retard evaporation to permit examination. If the cytoplasm is too blue, back down to 95% alcohol. If the blue has been lost from the chromosomes, carry back to water and IKI and restain.

The Tannic Acid-Ferric Chloride Stain (Foster)

This stain is used for meristematic tissues, in which it stains the thin cell walls. Because of the simplicity of the schedule, no chart is necessary. The reagents that bring about the staining are as follows:

1. Tannic acid, 1% aqueous, with 1% sodium benzoate as a preservative.
2. Ferric chloride, 3% aqueous solution.

The procedure from water is as follows:

1. Tannic acid 10 min.
2. Wash thoroughly in water.
3. Ferric chloride, 2–5 min.
4. Wash in water, and examine with microscope.

Repeat steps 1 to 4 inclusive until the cell walls are sharply outlined. Nuclei may be stained in safranin if desired, using aqueous safranin, or safranin in 50% alcohol.

Variations of the fundamental method are described by Northen (1936). This stain is likely to undergo further modification and will probably become one of our most useful histological stains.

The foregoing outline of the elements of staining processes is likely to be adequate for the average needs of students and teachers and for many research problems. The beginner is warned not to dabble in a wide variety of processes but to gain a mastery of a few fundamental methods. Ability to analyze and remedy difficulties should be cultivated. The advanced worker who finds that a research problem requires more specialized methods should turn to the literature and search out methods that have been used for similar investigations.

Some of the processes described in this chapter have been used in cytological research for many years. These methods have been replaced by smear and squash techniques, ultra-thin sectioning, and electron microscopy in studies that are concerned primarily with the structure and behavior of chromosomes within a cell. Nevertheless, the well-established "thick" sectioning and staining techniques are still indispensable for cytological studies in which it is imperative to keep intact the organization of tissues and organs. Some examples are the study of meristems, ovules, embryology, as well as tumors and other aberrations induced by chemical agents, irradiation, or invading organisms. An extensive list of stain formulas and procedures can be found in Gray (1954); some basic procedures are presented in Gray (1952); a comprehensive treatment of histochemical tests is given by Jensen (1962).

8. The Celloidin Method

Infiltration

The celloidin process is used for subjects that are tough, brittle, or friable (crumbly). Paraffin does not afford adequate support for sectioning such materials. One example of a subject for which the celloidin process must be used is a graft union (Fig. 13.9 *a*), in which the incompletely united members must be kept intact before and after sectioning. Another illustration is the preparation of pathological materials which may be in such disintegrated, fragile condition that the sections would fall apart without the celloidin matrix. Sectors from large trees having wood, cambium and bark tissues cannot be kept intact without embedding in celloidin (Fig. 13.9 *b, c*). Unembedded small twigs are difficult to hold and orient for longitudinal sectioning without some matrix. A twig or sector from a tree can be embedded in celloidin, blocked as shown in Fig. 8.2 *D–F,* and sectioned accurately in any desired plane. A properly selected piece of material may yield 100 sections, uniform in thickness, orientation, and staining properties. Permanent slides can thus be made by the hundreds at low cost. In some laboratories the celloidin method is neglected, or even scorned, but the above illustrations show that the process has its place in any well-equipped, versatile laboratory.

The matrix for the celloidin process is a form of nitrocellulose, known by several trade names — celloidin, collodion, Parlodion, and some less common names. This product is sold in the form of shreds or chips, packed dry or in distilled water. The latter method retards the development of a yellow color. Celloidin should be dried thoroughly before being dissolved for use. The most commonly used solvent consists of approximately equal volumes of ether and methyl alcohol. These reagents must be of the best quality and *anhydrous.* Five stock solutions of celloidin are usually used. These solutions contain, respectively, 2, 4, 6, 8, and 10 g. of dried celloidin per 100 cc. of solvent. These are designated for convenience as 2% celloidin, etc.

Infiltration in celloidin consists of transferring the previously killed and dehydrated tissues into a dilute solution of celloidin, concentrating the celloidin, and finally molding the thickened celloidin into blocks containing the material. Concentrating the matrix may be accomplished by one of the following processes or by combinations of processes.

1. Transfer the tissues thorugh a graded series of celloidin solutions of increasing concentration.

2. Add chips of dry celloidin at intervals to the initial 2% solution.

3. Evaporate the solvent from a large volume of a 2% solution.

METHOD 1

Transfer the dehydrated tissues from the dehydrant into the solvent. After one to several hr., transfer to 2% celloidin, covering the material with at least five times its volume of celloidin. Fasten a dry, rolled cork into the bottle by means of wire loops (Fig. 8.1 *A, B*). Put the bottle into the 53°C. oven.

The interval in the oven varies widely. For sections of twigs having a diameter of 3 to 5 mm., 24 hr. in 2% celloidin may be enough. For larger pieces and for dense materials increase the time to 2 days or more. After the interval in 2% celloidin, *cool the bottle,* remove the stopper, and pour the celloidin into a dry pan (not into the sink!). Keep away from flames or sparks. Cover the tissues immediately with 4% celloidin. Reseal the bottle, and repeat the interval under pressure in the oven. Repeat this operation with 6, 8, and 10% celloidin. Following the last treatment, continue to thicken the celloidin by adding a chip of dry celloidin every 24 hr. When the celloidin is so thick that it just flows at room temperature, the material is ready to be hardened as described on page 81. To determine whether the consistency of the celloidin is correct for blocking, dip a thoroughly dried matchstick into the celloidin, lift out a mass of celloidin, and immerse in chloroform for 1 hr. The celloidin should become hardened into a clear, firm mass that can be sliced easily with a razor blade. A comparison of samples taken during successive stages of infiltration will show the progressive increase in firmness and improved cutting properties.

METHOD 2

The material is first given at least 48 hr. in 2% celloidin in a sealed bottle in the oven. At intervals of several days the bottle is cooled and unsealed, a chip of dried celloidin added, and the bottle sealed and returned to the oven for the next interval. When working with delicate material, the celloidin chip should not be dropped onto the material but tied into a bag of dry cheesecloth, which is then suspended in the bottle so that the celloidin is just immersed in the solution (Fig. 8.1 C). The periodic addition of celloidin is continued

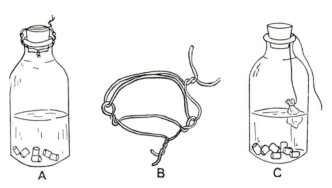

A   B   C

Fɪɢ. 8.1—*A*, Specimen bottle for infiltration with celloidin under pressure, with stopper fastened by wire loops; *B*, detail of wire loops for fastening cork; *C*, cheese-cloth-bag method of thickening celloidin.

until the solution is thickened to the degree described in the preceding method.

METHOD 3

This method is very slow but yields superior results. The material is started in a large volume of 2% celloidin, at least four times the depth occupied by the material. Mark the initial level of the solution. Cork the bottle loosely, but wire the cork so that it cannot be pushed out. Keep the bottle in a warm place, away from flames or sparks. Slow evaporation takes place, and, when the volume is one-half the original, the solution is in approximately 4% celloidin. Add new 4% celloidin to make up the original volume. If the celloidin has become colored, replace with new 4% solution. Continue the process of slow evaporation until the material is in thick celloidin. An objectionable feature of method 3 was pointed out by Walls (1936). If the evaporation rate is too rapid, it seems that the ether of the solvent evaporates more rapidly than the methyl alcohol, and the celloidin jells before adequate thickening is obtained. This can be remedied by adding a small quantity of pure ether and continuing infiltration until the proper viscosity is attained.

A low-viscosity nitrocellulose has been recommended for rapid infiltration of firm materials (Davenport and Swank 1934; Koneff and Lyons 1937). This inexpensive celloidin forms a firm matrix, and its solutions in ether-alcohol tolerate 6% water, thus minimizing the extreme brittleness produced in woody materials by total dehydration. The above references give complete details of procedure.

Hardening and Blocking

In the celloidin process the solvent is not eliminated completely during infiltration. The thickened celloidin solution is hardened by immersion in chloroform. Remove an infiltrated piece of material and a mass of enveloping celloidin and immerse in chloroform. The celloidin loses its stickiness at once and soon becomes hardened throughout. It is best to leave the material in chloroform for 12 hr. to harden the celloidin in the innermost cells of the material. Transfer the hardened material into a mixture of approximately equal volumes of 95% ethyl alcohol and glycerin, in which the material may be stored indefinitely.

Large pieces of embedded wood may be removed from the glycerin-alcohol, clamped directly into the microtome, and sectioned. Subjects having easily separable soft tissues are often damaged by compression in the clamp. Such materials are sectioned best by mounting them on blocks of wood or plastic. The mounting block may then be clamped rigidly into the microtome clamp without damaging the tissues. A twig or other long slender object should be mounted into a plastic tube or a wood block having a hole of suitable size drilled lengthwise through the mounting block (Fig. 8.2 A–C). Prepare mounting blocks by drying them thoroughly in a 110°C. oven, soak in anhydrous methyl alcohol, then store in waste 4% celloidin until needed. When the material being infiltrated is put into 8% celloidin, drop a prepared mounting block into the specimen bottle and continue the infiltration.

To mount twigs for cutting transverse sections, remove the desired twig and a suitable drilled mounting block from the thickened celloidin, and push the twig into the hole, leaving 6 to 10 mm. of the twig protruding. Fill any space remaining around the twig by pushing slivers of matchstick into the hole from below. Wrap a generous mass of thick celloidin around the twig and mounting block, and harden in chloroform (Fig. 8.2 C). For longitudinal sections of a twig, lay the infiltrated twig on a large, undrilled block, wrap well with additional thick celloidin, and harden in chloroform.

When the surface of the celloidin is hard (2 min.), press the twig gently until it is flat on the mounting block, thus affording firmer support for sectioning. A batch of embedded material usually contains more pieces than are needed for immediate sectioning, therefore, only a few pieces need to be mounted on blocks. Most of the pieces are merely removed from the thick celloidin, hardened in chloroform, and stored in glycerin-alcohol. The pieces can be blocked at any future time by the method to be described later.

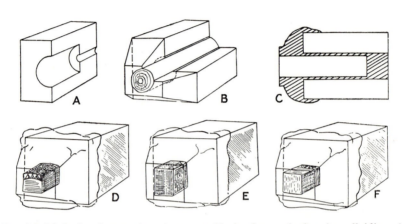

FIG. 8.2—Methods of mounting tissues on blocks for sectioning in celloidin: *A*, cutaway view of drilled block for holding long object; *B* and *C*, cutaway views of drilled block containing twig or other long object enveloped in hardened celloidin; *D–F*, infiltrated blocks of wood and bark embedded in hardened celloidin, oriented on mounting blocks to cut transverse, radial, and tangential sections, respectively.

Sectors from large limbs usually must be fastened on mounting blocks for sectioning. If the pieces have cambium and other tissues of the bark, these tissues may peel off when the piece is compressed in the microtome clamp. Mount three pieces from each subject, on separate blocks, so that transverse, radial, and tangential sections can be cut (Fig. 8.2 *D–F*). A generous wrapping of celloidin should envelop part of the mounting block as in Fig. 8.2. The rigidity of the mounting of such material can be improved by mounting the material in a recess that has been drilled about 1/16 in. deep into the end of the mounting block.

Blocking of previously hardened embedded material is a simple operation. Remove the desired pieces from the glycerin-alcohol storage fluid and soak in anhydrous *ethyl* alcohol. Change the alcohol twice at 4- to 8-hr. intervals. This removes the small amount of water left from the storage fluid, softens the celloidin, but does not dissolve an appreciable amount. Transfer the pieces into thick celloidin of a consistency suitable for casting. Also put a supply of mounting blocks into this thick celloidin. After at least 24 hr. in thick celloidin, mount and harden as previously described. It is sometimes necessary to trim pieces of tissue prior to reblocking in order to establish the correct cutting planes. Trimming should be done when the pieces are removed from glycerin-alcohol. The glycerin prevents drying and shrinkage of the tissues during trimming.

Waste celloidin from various stages in the process can be salvaged by evaporating in a shallow pan in a place free from dust and open flames. The dried sheet is cut into shreds, dried at 53°C., and used to make solution for treating mounting blocks and to make 2 and 4% celloidin for method 1. Celloidin solution that is too discolored to be salvaged is most easily disposed of by pouring it into a pan of cold water. The celloidin hardens into a crust which can be lifted out and discarded.

Cellosolve is the trade name used for two synthetic organic compounds, ethylene-glycol-monoethyl ether and its methyl homologue. These fluids are solvents of celloidin and may ultimately replace the inflammable alcohol-ether solvent used heretofore. These solvents are not inflammable at ordinary working temperatures, therefore the entire process may be carried out in open or loosely stoppered bottles. The evaporation rate is very slow at 50 to 55°C. Methyl Cellosolve, which boils at 124.3°C., evaporates slightly faster than ethyl Cellosolve, which boils at 135.1°C. The latter solvent is preferred for fragile subjects which tend to collapse if the celloidin concentration is increased too rapidly.

An inexpensive method of using Cellosolve is to dehydrate the tissues in the appropriate grades of alcohol and to transfer to a 2% solution of celloidin in Cellosolve. Subsequent infiltration may be accomplished by successive treatment in 4, 6, 8, and 10% solutions at 50 to 55°C., or by beginning with 2% and periodically adding celloidin chips by the cheesecloth-bag method. The interval in each grade ranges from 24 hr. for small or porous pieces to a week for large blocks of wood.

Cellosolve may also be used as the dehydrating agent. Materials in which the preservation of the protoplast is not important may be transferred, after killing and washing, directly into Cellosolve. Make two or three changes of Cellosolve before beginning the infiltration. However, if the material requires thorough or gradual dehydration to insure adhesion of the celloidin, it is better to dehydrate in ethyl alcohol or acetone and to use the much more expensive Cellosolve as the infiltration solvent.

Special Treatment of Hard Woods

The foregoing methods of infiltration yield excellent results with some soft woods such as willow, poplar, basswood, white pine, and many other woods. These can be infiltrated in celloidin without special preliminary treatment, but oak, hickory, walnut, the yellow pines, and other woods are too hard to section by the regular process. Such materials can be softened by treating with hydrofluoric acid (HF). This highly corrosive reagent is purchased in wax bottles and should be used in wax or wax-lined or plastic containers. Because of the corrosive action of the liquid and vapor on glass, metals, and the skin, HF should be used in an isolated part of the premises, away from valuable instruments. The staining and microchemical reactions of tissues are materially altered by this treatment.

Twigs having living bark tissues are first killed as usual and transferred to HF. Dry woods are prepared for treatment in HF by alternate boiling in water and exhausting in an aspirator in cold water until the pieces are saturated. The safest concentration of HF for most subjects is commercial acid diluted with approximately twice its volume of water. The duration of treatment in HF varies greatly with the hardness of the material, the size of the pieces, and other factors. As a trial, treat a hard wood such as oak for 5 days, wash in running water for at least 1 hr. to make it safe to handle the pieces, and try to cut thin slices with a sharp razor blade. An alternative method of testing is to wash the pieces in running water for 4 hr., clamp a piece into the sliding microtome, and test its cutting properties. After making a test, wash and wipe the clamp and the knife thoroughly. If the material is too hard to cut readily either freehand or with the microtome, return to the HF for another 3- to 5-day interval, and test again. When the wood seems to cut satisfactorily, wash for at least 48 hr., whether it is to be embedded in celloidin or cut without embedding. Wood or twigs that have been treated with HF and are to be put into storage without embedding should be dehydrated through 20, 40, and 60% alcohol or acetone at 4- to 8-hr. intervals, then stored in a mixture of equal volumes of alcohol, glycerin, and water.

The softening of wood can be accelerated by treating in HF under pressure. The necessary equipment is not available commercially and must be built to specifications. A satisfactory apparatus, described by Chowdhury (1934), consists of a section of iron pipe with a threaded flange at each end. Plates are bolted to the flanges, and the upper plate is removable for introducing the specimens. The compression chamber is lead lined and is provided with a pressure gauge and a valve to which the pump is attached. Chowdhury recommends 40% HF and a pressure of 80 lb. He found that 1-in. cube blocks of *Juglans regia* were adequately softened in 3 days; blocks of *Diospyros melanoxylon,* an extremely hard wood related to our persimmon, required 7 days. The equipment necessary for this method is amply justified if a considerable amount of diagnostic work on timber woods is being carried on.

Material embedded in celloidin can be cut as soon as the celloidin has been hardened in chloroform and the volatile chloroform replaced with the glycerin-alcohol storage fluid. The cutting properties are improved by prolonged storage in glycerin-alcohol. If materials having dark-colored bark and light-colored wood are stored for several years, the storage fluid dissolves coloring matter from the bark and imparts a dark color to the wood. Stained sections from such wood do not have bright, clear colors. The stock of embedded twigs of basswood, for example, should be replaced every 3 to 5 years. Incomplete removal of killing fluids or of hydrofluoric acid results in gradual disintegration of stored material.

Sectioning

Celloidin sections are usually cut with a sliding microtome. In this type of instrument the material is stationary during the cutting stroke, while the knife carriage slides on an accurate track. An automatic or hand-operated feed mechanism moves the tissues upward between cutting strokes. The catalogues of the leading manufacturers contain instructive illustrations and descriptions of several types and price classes of sliding microtomes. Small pieces of moderately soft tissues can be cut with a razor blade in a special holder designed for the sliding microtome. The limitations of the razor blade must be determined by trial. Hard materials and large sections must be cut with a microtome knife. Various lengths and weights of knives are available. The method of sharpening a microtome knife is described in Chap. 6.

Before using the sliding microtome, wipe the track of the knife carriage with an oiled cloth and test the feed mechanism. Clamp the knife firmly into the sliding carriage. Remove a piece of blocked tissue

(Fig. 8.2), fasten into the microtome clamp, and adjust the universal joint until the desired plane of sectioning is parallel to the plane of travel of the knife (Fig. 6.3). Keep the tissues moistened with glycerin-alcohol. If the upper surface of the material is not level, trim with a razor blade, sparing the microtome knife from rough trimming work. The best cutting angle for the knife-edge, with reference to the line of travel, ranges from 30 to 40°. The vertical tilt or declination of the flat side of the knife is also subject to variation. Begin with just enough tilt to enable the back of the ground wedge to clear the tissues (Fig. 6.3 *A, B*). Bring the tissues into cutting contact with the knife, using the hand-operated feed, making each vertical feed movement *after* the knife has passed over the material on the *return stroke*. In order to avoid damaging the knife-edge, feed in steps of not over 15 μ. Make sure that there will be ample clearance between the knife carriage and the tissue carrier even after many sections have been cut.

When each stroke cuts a complete section, set the thickness gauge and the automatic feed device. A thickness of 15 to 20 μ is satisfactory for most woody subjects. Keep the material and the knife flooded with 95% alcohol while cutting sections, and transfer each section as soon as it is cut to 95% alcohol in a watch glass or other shallow container. Vary the cutting angle and declination until sections slide up onto the knife without compression, curling, or breaking. Newly embedded material is liable to be hard and brittle and to curl. Curling can be minimized by holding a finger, moistened with alcohol, in light contact with the material during the cutting stroke, until the knife has cut through the marginal celloidin and enters the material. If scratches are evident on the cut surface, the portion of the knife being used may have bad nicks. Shift the knife longitudinally in its clamp and discard the next few sections. Sections can be stained at once, or they may be stored in glycerin-alcohol indefinitely. In the case of materials that do not curl, it is possible to cut several hundred sections, store them in a bottle of glycerin-alcohol, and remove as many as needed for staining at any time.

Some materials can be cut readily enough but difficulties arise after sectioning. The sections may curl soon after removal from the knife and become increasingly tightly curled during staining and dehydration. Being made brittle by dehydration and clearing, the sections break when an attempt is made to uncurl them for mounting. The following method is usually effective for such material. Keeping the knife well flooded with alcohol, cut a section. Hold a finger under the knife-edge and float the section onto the finger, with the concave side of the section upward. Press the section, with the concave side *down,* on a slide flooded with a thin film of glycerin-alcohol. Line up successive sections on the slide, where they lie flat; drying of the sections is prevented by the glycerin. When enough sections have been cut, press a dry slide over the sections. Transfer the slides with the sections pressed between them to a dry Petri dish, put lead weights on the top slide and fill the dish with water. As many as four pressed lots can be put into one Petri dish. The water renders the sections flexible and permits them to flatten. The sections are then floated out and stained. For prolonged storage, transfer the slides and weights to a Petri dish of glycerin-alcohol, in which the sections become hardened in a flattened condition and in which they may be kept pressed indefinitely. Sections that have been stored either floating or pressed in glycerin-alcohol are progressively transferred to water and stained.

Staining

Sections cut in celloidin on the sliding microtome do not adhere to form a ribbon. They are usually stained as loose sections floating in a watch glass or small evaporating dish. The sections are usually floated off the knife into 95% alcohol. For staining in an aqueous stain, sections are gradually transferred to water. As the first step, add about one-third as much water as there is alcohol. After 3 to 5 min. pour off half of the liquid, and add an equal volume of water. Repeat the decantation and addition of water two or three times, then drain off all the liquid, and rinse in water. From this point the sequence of operations conforms in general to Staining Chart III. Drain and cover with hemalum. After 5 min. in stain remove a section with a brush, rinse in distilled water, then in tap water, and examine with a microscope. The intensity of hemalum is correct when the cambium, phloem, cortex, pith, and xylem rays are blue, but lignified tissues are practically colorless. Nuclei should be blue-black. Drain off the stain, and rinse the sections in three to five changes of distilled water and two or three changes of tap water. Overstained sections can be destained in ½% HCl, followed by thorough washing in tap water. When the intensity of the blue color is correct, cover the sections with aqueous safranin. Some woody materials take up enough safranin in 5 to 10 min. Materials having less highly lignified cell walls may require 12 hr. After the estimated time in safranin, rinse with water until the rinse water is colorless. Flood with 50% alcohol, in which destaining of safranin begins to take place. Do not use acetone until the anhydrous stage. After 3 to 5 min., change to 95% alcohol, in which destaining continues. At first the blue color of the hematoxylin is completely masked by the red safranin, but as destaining proceeds the blue color becomes evident. At intervals of 2 to 5 min. transfer a section to a watch glass of clean 95% alcohol, and examine with a microscope. When there is good contrast between the blue color of nonlignified tissues and the clear, brilliant red of lignified elements, drain, and rinse in five changes of anhydrous alcohol or acetone at 2- to 5-min. intervals. Drying of the sections must be avoided in making these changes. Destaining is almost entirely stopped in anhydrous acetone, and the celloidin matrix is dissolved out of the tissues. If absolute alcohol is used, make two changes of ether-alcohol solvent to dissolve the celloidin out of the tissues. Flood with fresh carbol-xylene. There should be practically no destaining action during the 5- to 10-min. interval in carbol-xylene. Rinse in five changes of xylene, and mount in balsam or synthetic resin.

If the celloidin support is dissolved out of some pathological materials, graft unions, and some fragile subjects, the sections disintegrate or lose important parts. The sections can be cleared by transferring directly from 95% alcohol to terpineol, carbol-xylene, or creosote. The clearing agent must be changed several times, and thoroughly rinsed out in xylene before mounting. The supporting celloidin is retained by this method.

To mount one section on each slide, remove a section from xylene with a small brush or section lifter and place the section on the center of a dry slide. Stained sections can be selected under a binocular microscope or a hand lens and the imperfect sections discarded. Keep the section on the slide moistened with xylene. If the section is curled, straighten with two brushes, keeping the concave side down. Place a drop of resin on the section and lower the cover glass obliquely, squeezing out air bubbles by gentle pressure or by tapping with the eraser on a pencil. Put a lead weight on the cover glass. The drop of resin should be of such size that there is no excess resin around or over the weighted cover glass. If air bubbles cannot be expelled or if too much resin was used, put the slide into a Petri dish of xylene. The cover glass and section can be slid off in a few minutes and the section remounted. It is much easier to uncover and re-cover than to clean up a messy preparation later. After 1 to 3 days of drying under pressure in a horizontal position, the weight may be removed, and the slides labeled and boxed. Refer to Fig. 6.6 for suggestions on the selection of cover glasses of suitable size.

The foregoing method is rapid, highly productive, and entirely satisfactory with materials that do not curl during the staining process. It is often possible to stain at least 50 sections in a Petri dish or evaporating dish and to mount most of the sections before appreciable brittleness and curling develop. Sections that undergo rapid curling after removal of the matrix must be handled by other methods. Terpineol has the valuable property of clearing stained and dehydrated sections without making them brittle and without affecting the stain. The terpineol is introduced in place of carbol-xylene. Sections may be lifted singly from terpineol, rinsed in xylene, and promptly mounted. If sections curl with this method, remove the sections singly from terpineol and place them lined up in rows on a dry slide until the slide is filled. Place the sections with the concave side down, and keep them moistened with terpineol. Cover with another dry slide. Place the slides with the pressed sections into a dry Petri dish, and flood with xylene. After 4 to 8 hr. the sections will become hardened flat, and they can be floated out a few at a time, rinsed in two changes of xylene, and mounted in resin before serious curling occurs.

The foregoing staining process, using a self-mordanting hematoxylin and safranin, is but one of the many stain combinations that can be used for celloidin sections. Safranin is almost invariably one component, because of the highly lignified character of most plant subjects for which the celloidin method is used. The safranin and fast green combinations yield strikingly beautiful preparations with many subjects. A batch of blue ash stem, killed in *FAA* showed highly differential, clear and brilliant tones, whereas a batch of red elm, similarly processed, had an unattractive, hazy blue tone in tissues that should have stained green. This stain combination can be tested rapidly with a few sections from any subject and deserves a trial. Follow the sequence given in Staining Chart V, observing the precautions and modifications necessary with celloidin sections.

Iron hematoxylin is an important stain for celloidin sections, because of its sharp selectivity for the middle lamella. The sequence is the same as in Staining Chart VI, but the time in mordant and stain, respectively, need not exceed 1 hr. After the stain has been differentiated, washing in water must be very thorough because woody tissues retain the alum tenaciously, resulting in early fading of the stain in the finished preparation.

Celloidin-Paraffin Double Embedding

Double embedding consists of infiltrating and embedding tissues in celloidin and then infiltrating with paraffin. This procedure is used with materials that combine hard tissues with regions of very fragile and brittle tissues. The stems of some grasses and sedges do not become well infiltrated by celloidin, but paraffin penetrates well. The material has regions of highly lignified sclerenchyma, requiring more support than that afforded by paraffin alone.

Embed in celloidin by one of the foregoing processes and harden well in chloroform. Trim away the enveloping celloidin, exposing all cut surfaces but leaving intact the outer surface of the epidermis. Some workers use clearing oils or mixtures that clear or make the tissues transparent. This does not necessarily improve the subsequent infiltration and usually aggravates brittleness. It is adequate to change the chloroform several times to eliminate the celloidin solvent and to proceed with infiltration in paraffin. The embedded material may be cut and ribboned on a rotary microtome; very firm material must be cut on the sliding microtome.

Rapid progress is being made in the use of methacrylates, polyesthers, and various polymers for embedding, especially for ultra-thin sectioning (Newman, Borsyko, and Swerdlow, 1949; Massey, 1953; Kuhn and Lutz, 1958). The use of such materials as additives to embedding wax is being explored (see page 33). It is possible that these new matrices may replace to some extent the use of celloidin for hard tissues.

9. Sectioning Unembedded Tissues

Materials that have sufficient rigidity to withstand the impact of the sectioning knife can be sectioned without embedding. The most rapid and simple method consists of grasping a piece of fresh or preserved plant material in the fingers and slicing with a razor or a razor blade. Extremely thin sections can be cut in this way by a skilled and experienced worker. These methods receive entirely too little attention in teaching and research. These seemingly crude methods can yield excellent preparations for teaching. The student who collects his own materials and makes his own preparations, even though crude ones, gains an understanding of plant structure that cannot be imparted solely by thrusting a neatly labeled finished slide before him. Much wasted effort could be spared in research by adequate preliminary survey work conducted by freehand sectioning.

Written directions are of little instructional value for this work. Patience, experience, and perhaps inherent skill are the chief requirements. Sectioning can be aided by enclosing the material between pieces of pith or cork. Split a cylinder of pith lengthwise and cut a longitudinal groove or a recess in the pith of appropriate size and shape to receive the specimen. Wrap the two pieces of pith together with thread, and soak in water. The pith expands and encloses the material firmly enough to be sliced with a sharp blade. The tissues and the knife should be kept wet with water and the sections floated in water. The subsequent handling of the sections is described at the end of the discussion of sectioning. A drip siphon is a useful device when doing extensive freehand sectioning. Place a 2- to 5-liter bottle of distilled water above the work table, and install a siphon that terminates in a glass tube drawn to a fine aperture. Adjust the siphon with a screw clamp so that a drop of water is released every 2 or 3 sec. and drops into a waste container. The worker can have both hands occupied with the sectioning, and a drop of water for wetting the material or floating a section is instantly available at any time.

Sectioning Unembedded Tissues With the Microtome

Rigid materials or large objects from which it is difficult to obtain complete sections freehand should be cut with a microtome. Any sliding microtome, from the inexpensive table microtome to the most elaborate precision microtome, may be used. A fresh twig of white pine, basswood, or cottonwood makes an excellent subject. Clamp a 3-cm. length of twig into the microtome with 1 cm. projecting above the clamp. Set the knife so that it makes a long slanting cut, and make sure that there will be ample clearance between the tissue clamp and the knife carriage. Keep the twig and the knife flooded with water while cutting. Remove the slices from the knife, and float them in a watch glass of water. Examine the floating sections with a binocular or a hand lens, discard imperfect sections, and continue sectioning and selecting until enough satisfactory sections have been cut. The ratio of perfect sections obtainable by this method is much less than is possible by the celloidin method. Nevertheless, the quality and output by sectioning fresh material are an agreeable surprise to workers who give it a fair trial.

The above method can be used with unembedded tissues that have been killed in 70% alcohol and stored in that fluid, or with materials that have been killed in any fixing fluid and dehydrated to 70% alcohol. This degree of dehydration is usually necessary to make the tissues sufficiently firm. However, woody tissues may be killed in *FAA,* rinsed in several changes of 50% alcohol to eliminate much of the acid, and sectioned as above. When cutting tissues from a wet preservative or when cutting partly dehydrated tissues, keep the tissues and the knife flooded with water or with dilute alcohol of approximately the same water concentration as the preservative. Float the sections in a dish of the fluid in which they were cut.

Sections of living tissues frequently do not have satisfactory staining reactions, especially with the stains used for permanent slides. Furthermore, these staining processes usually induce severe plasmolysis and alteration of the protoplasm. If good preservation of the protoplasm is desired, transfer sections cut from living material into a killing fluid, in which the protoplasm is fixed and hardened. For sections of woody twigs or firm herbaceous stems, *FAA* is recommended. For critical studies try Craf II or III (Chap. 3) or an experimentally determined modification. The quality of preservation produced by a formula can be ascertained by examining sections in a drop of the fluid. Thin sections are killed and hardened almost instantly. After 10 min. in the fluid, rinse the sections in water, and proceed with the staining.

The preceding methods permit the study of living cells, or cells in which protoplasmic details were fixed by reagents. If these methods of sectioning unembedded material fail to give satisfactory results, the materials must be treated by methods which may distort or destroy fine protoplasmic details. Nevertheless, the following methods are useful, within the stated limitations.

Dry lumber of soft woods such as white pine, basswood, or willow can be sectioned successfully without embeddding. Trim the wood carefully into blocks measuring approximately 1 by 1 by 2 cm., and prepare the blocks by alternately boiling in water and pumping in cold water, until the pieces are thoroughly saturated and sink. Hard woods that cannot be cut after this treatment may be sectioned by one of the following methods.

The live-steam method of sectioning wood is based on the principle used in the manufacture of veneer. If a jet of superheated steam is directed upon the surface of a block of wood, the surface becomes soft enough to permit the cutting of a thin section. Steam can be generated in a flask, but a safer device is a small steam generator, provided with a pressure gauge, water level gauge, safety valve, and a water inlet. Such generators are obtainable from dealers in chemical apparatus. The steam from the generator passes through a copper tube, in which there is a superheating coil heated by a Bunsen burner, and the superheated steam emerges through the small orifice of a nozzle which can be adjusted over the specimen. Successive sections are cut at experimentally determined intervals of steaming. For additional details and variations of the apparatus refer to Crowell (1930) and Davis and Stover (1936).

Jeffrey's vulcanization method makes possible the sectioning of extremely hard materials, such as walnut shells. Materials are sealed in a chamber made from a section of pipe, and heated in a dental vulcanizer at approximately 160°C. for 1 to 5 hr. This is followed by treatment in hydrofluoric acid. Materials may be cut unembedded or in celloidin. Because of the special equipment necessary, the procedure is not described here, and interested workers are referred to Jeffrey (1928).

Sectioning With the Freezing Microtome

Unembedded tissues that are too soft or fragile to stand up under the impact of the knife can in some cases be frozen and then sectioned. The device used on the sliding microtome for this purpose is known as a freezing attachment. The various freezing attachments are described in the catalogues of the manufacturers of microtomes. The device replaces the usual tissue clamp or object carrier of the microtome. Freezing is accomplished by the evaporation or expansion of a freezing agent — ether, CO_2 gas, or solid CO_2 (dry ice). The piece of tissue to be frozen is usually enveloped in a fluid or semifluid medium, which on freezing affords additional support.

Gum arabic is probably the best-known supporting medium. Make an aqueous solution of gum arabic of thick, sirupy consistency. Add a few crystals of carbolic acid. Dip the specimen to be cut into the gum. Freeze a 2- to 3-mm. layer of gum on the supporting disk of the freezer. Place the specimen on the disk, wrapping a generous quantity of the gum around the specimen. Turn on the freezer, and, as the gum begins to congeal, wrap more gum on the specimen until the material is well supported. Proceed with the sectioning.

Gelatin is another satisfactory supporting medium. Make a gelatin solution that is semifluid at room temperatures. Add 0.1 per cent carbolic acid as a preservative. Warm on a water bath for use, and use in the manner described above for gum arabic. A semifluid solution of agar is another excellent medium, used like gelatin.

Freehand or microtome sections can be mounted in a drop of water or 50% glycerin and studied with the microscope. If glycerin is used, the water can be evaporated and the cover glass sealed with lacquer or paraffin, making a semipermanent mount. The mounting media described on page 102 also can be used. Sections of dark-colored woods, or other materials having adequate coloration, can thus be made into semipermanent or permanent slides without staining.

Staining

The staining of sections of unembedded tissues is essentially the same as the staining of celloidin sections. The worker who has had previous experience with paraffin sections can follow the staining charts in Chap. 7, making the necessary modifications. For the benefit of workers who wish to use the present chapter without having had previous experience, an outline of some simple, practical processes is offered. Sections cut in alcohol should be progressively transferred or run down to water before staining. Add an equal volume of water to the alcohol containing the sections. Mix gently, pour off half of the liquid, and add an equal volume of water. Pour off all the liquid, and rinse the sections with two changes of water. Proceed with the staining process.

The most easily controlled stain combination consists of a self-mordanting hematoxylin (Chap. 7) followed by an aniline dye. Assume that sections of a firm woody subject are to be stained. Drain off the water in which the sections are floating in a watch glass, and flood the sections with hemalum or similar stain. After 5 min. remove a section with a brush, rinse in distilled water and then in tap water and examine with a microscope. Lignified cells, such as tracheids and phloem sclerenchyma, should be practically colorless. Soft tissues like pith, cortex, and cambium should have blue-stained walls. Nuclei should be blue-black. If the sections are overstained the entire protoplasts become blackened, obscuring cellular detail, and the walls of lignified cells become stained. Sections can be destained with ½% HCl and thoroughly washed in tap water. When the correct intensity of blue is attained, cover the sections with safranin. Try an interval of 15 min. in safranin. Locate this point in the staining schedule given in Staining Chart III. Rinse the sections in water, and cover with 50% alcohol. The safranin will dissolve out of the nonlignified tissues faster than out of lignified cell walls. Slower destaining can be obtained with 50% acetone. When the lignified cells are still deep red, rinse the sections quickly in anhydrous alcohol or acetone. Acetone stops destaining action better than does alcohol. Complete the process as shown in Staining Chart III. Mount the sections on slides as described in Chap. 7.

Having gained some familiarity with the above stain and with the use of staining schedules, study the discussion of staining in the paraffin method (Chap. 7) and the celloidin method described earlier in this chapter, and try some of other stain combinations described in those chapters.

Sectioning by Encasing in Water-Soluble Waxes

These little-known methods are intermediate between embedding methods and sectioning without embedding. As a matrix, use one of the water-soluble wax-like synthetics, such as glycerol monostearate, which melts at 55°C. The living or fixed material is transferred directly from water to the melted matrix, which is then hardened, fastened to a wood mounting block, and sectioned. Very little infiltration occurs, but the material is encased and held with sufficient rigidity to make fairly good sections. This method has been used successfully with nearly mature clover seeds and shows much promise for similar subjects. Other synthetic stearates should be tried (Johansen 1940, McLane 1951).

Microchemical Tests

One of the important advantages of sectioning fresh untreated material is the avoidance of the chemical and physical alterations that are undoubtedly produced by the processing necessary for embedding. Although the protoplasm is probably changed by the handling incident to sectioning and mounting fresh material, the nonliving constituents of cells and tissues probably are not markedly changed chemically. This makes possible the use of microchemical tests that reveal with more or less accuracy the chemical nature of important structures. Although the science of chemical microscopy is highly developed, it occupies a minor place or is virtually ignored in many botanical curricula. However, certain chemical tests are generally regarded as indispensable in even an elementary study and are therefore included here.

Starch; Iodine-Potassium Iodide Test (IKI)

Water .... 100 cc.
Potassium iodide .... 1 g.
Iodine .... 1 g.

Place a drop of the reagent directly upon the specimen. Most starches give a blue-black color. Waxy starch, found in some genetic stocks of maize, turns yellow or brown. By using a very dilute solution of the reagent and imparting only a trace of color to the starch, the laminations in the granules may be observed with the microscope. When testing entire living cells such as those of *Spirogyra* or leaves like those of *Elodea,* the aqueous reagent reacts very slowly, and a reagent made with 70% alcohol should be used.

Sugars; Osazone Test

Solution A

Glycerin (warm) .... 10 cc.
Phenylhydrazine-hydrochloride ... 1.0 g.

Solution B

Glycerin .... 10 cc.
Sodium acetate................. 1.0 g.

Mix a drop of each solution on a slide, float the sections in the mixture, place the slide over the mouth of a wide-mouthed flask containing boiling water, and heat for 10 to 15 min. Glucose and fructose produce fascicles of yellowish needles; maltose produces fan-shaped clusters of flattened needles. After 30 to 60 min. of heating, sucrose becomes hydrolyzed and reacts to form needles like those produced by glucose.

Reducing Sugars; Fehling's Solution Test

Although this is not a microtest, it is included because it is essential for a systematic examination of the prominent chemical constituents of cells and tissues.

Solution A

Water	1 l.
Copper sulphate	79.28 g.

Solution B

Water	1 l.
Sodium potassium tartrate	346 g.
Sodium hydroxide	100 g.

Mix equal volumes of *A* and *B* in a test tube, add a quantity of the finely pulverized materials to be tested, heat to boiling. A brick-red precipitate indicates reducing sugars.

If a negative or slight test is obtained for reducing sugars, a test for sucrose can be made by first hydrolyzing the sucrose. Add 1 cc. concentrated HCl to 10 cc. of the extract to be tested. Heat in a water bath at 70°C. for 5 min. Cool and neutralize with sodium carbonate, and test for the resulting reducing sugar with Fehling solution.

Lignin; Phloroglucin Test

Solution A

Phloroglucin, 1% to 2% in 95% alcohol

Solution B

Hydrochloric acid (try concentrated acid, as well as acid diluted with 1 to 3 volumes of water.)

Float the sections in a drop of phloroglucin on a slide, and cover with a cover glass. Place a small drop of the acid at one edge of the cover glass. Examine with a microscope. Lignified walls become violet-red.

Cellulose; Iodine-Sulphuric Acid Test

Mount sections or crushed fragments in IKI. Observe with the microscope, and locate blue-stained starch. Place a drop of 75 per cent H_2SO_4 at one side of the cover glass. As the acid diffuses in, note that cellulose walls swell and become blue.

Cellulose; Chloriodide of Zinc Test

Water	14 cc.
Zinc chloride	30 g.
Potassium iodide	5 g.
Iodine	0.9 g.

Mount thin sections in a drop of the reagent. Cellulose becomes blue.

Proteins; Millon's Reagent Test

Concentrated nitric acid	9 cc.
Mercury	1 g.

When dissolved, dilute with an equal volume of water. Place the specimen on a slide, drain or blot off excess water, put on just enough reagent to cover the material, and heat with a small flame. Proteins give a brick-red color. This is not a highly satisfactory reagent. Futhermore, it is highly corrosive and must be used with care. Do not permit inexperienced students to use this reagent on a microscope. The instructor should set up a demonstration microscope, after draining excess reagent from under the cover glass.

Fats and Oils; Sudan III

Alcohol (80%) .... 100 cc.
Sudan III .... 0.5 g.

Cut very thin sections or smear a fragment of the material on a slide, flood with the dye, and cover with a cover glass. After 10 to 20 min. the microscopic globules of fat should assume the bright, clear color of the dye. Cotyledons of the soybean and peanut are good subjects.

10. The Preparation of Whole Mounts and Smears

Preparations made without sectioning may be roughly classified as follows:

1. Dry preparations: herbarium sheets, Riker mounts, and bulk specimens. These methods are not within the scope of this manual.

2. Wet preparations: museum-jar preparations; bulk material for dissection.

3. Whole mounts, smears, and macerations for microscopic study.

Wet preservation may be used for a wide range of subjects in all major categories of the plant kingdom. Many subjects, such as the algae, can be preserved for critical study only by wet preservation. Entire plants or parts such as leaves, flowers, and fruits can be preserved in fluids that kill the cells, prevent decay, preserve the material in firm condition, and possibly preserve the natural colors.

The best-known preserving fluid is ethyl alcohol, usually used at a concentration of approximately 70%. This fluid preserves even the bulkiest objects. Considerable brittleness is produced, but preserved material can be made flexible for dissection by soaking it in water. Cells are plasmolyzed by alcohol, but this is not objectionable in dissections or freehand sections, in which the condition of the protoplasm is not important. Ethyl alcohol is difficult to obtain in some institutions, its use at field stations may be undesirable, and shipment of materials in alcohol involves legal technicalities. Regardless of these objections, ethyl alcohol is firmly established as a preserving fluid. Isopropyl alcohol may also be used.

Formaldehyde is an excellent preservative. This reagent is obtained as an aqueous solution containing 37 to 40% formaldehyde gas by weight. The U.S.P. (United States Pharmacopoeia) grade is adequate for preserving bulk materials. The most useful concentration for bulk preservation contains 5 parts formaldehyde solution in 95 parts of water. For massive objects the concentration must be doubled. At these concentrations the material does not become brittle, and some materials become pulpy after prolonged storage. Formaldehyde vapor and the solution are highly irritating and poisonous, producing persistent skin and pulmonary disorders. Materials that have been stored in strong formaldehyde should be rinsed in water if used for prolonged study.

An improvement over formaldehyde for the preservation of algae is the following:

Water .... 93 cc.
Formaldehyde (U.S.P.) .... 5 cc.
Glacial acetic acid .... 2 cc.

Hydrodictyon and *Spirogyra* stored in this fluid for 5 years were found to be in excellent condition for whole mounts in water. A further improvement is to add glycerin.

Water .... 72 cc.
Formaldehyde (U.S.P.) .... 5 cc.
Glacial acetic acid .... 3 cc.
Glycerin .... 20 cc.

This is one of the best preservatives for unicellular, filamentous, and even the larger bulky algae, for fleshy fungi, for liverworts, and for mosses. A trace of fast green dye imparts enough color to minute or transparent subjects to make them more readily visible under the microscope. The material should be mounted on a slide in a drop of the preservative. The volatile ingredients soon evaporate, but the glycerin prevents drying of the preparation during a long period of study. Liverwort thalli can be removed from the fluid and placed in a watch glass for study. The thalli of *Marchantia,* for example, are so firm that the gametangial disks stand upright in a normal position. The glycerin keeps the material moist for hours, and specimens can be returned to the stock bottle uninjured.

The natural colors of plants can be preserved in one of several formulas. The simplest formula for green plants consists of one of the *FAA* formulas with copper sulphate added (Blaydes 1937). The following formula is usually satisfactory:

Water .... 35 cc.
Copper sulphate .... 0.2 g.

When completely dissolved, add

Glacial acetic acid .... 5 cc.
Formaldehyde (U.S.P.) .... 10 cc.
Ethyl alcohol (95%) .... 50 cc.

Exhausting the air from the submerged specimens with an aspirator aids penetration. If the materials can withstand brief boiling in the fluid on a water bath, penetration and fixation of the color are hastened. Materials preserved in this formula can be subsequently embedded and sectioned, provided that the pieces are small enough to insure quick penetration of the preservative and satisfactory preservation of cellular details.

Keefe's formula is one of the best and should be used if the expensive uranium salt is available.

    50% alcohol .... 90 cc.
    Formaldehyde (U.S.P.) .... 5 cc.
    Glycerin .... 2.5 cc.
    Glacial acetic acid .... 2.5 cc.
    Copper chloride .... 10 g.
    Uranium nitrate .... 1.5 g.

Delicate subjects may be ready to use in 48 hr., but most materials require 3 to 10 days for complete fixation of the color. Leafy plants can be treated and then mounted as herbarium specimens, in which the color will persist for many months. This formula does not preserve the colors of flowers, nor is it satisfactory for gymnosperms.

The red and yellow coloration of fruits can be preserved in the following formula (Hessler) :

    Water .... 1 l.
    Zinc chloride (dissolve in boiling water and filter) .. 50 g.
    Formaldehyde (U.S.P.) .... 25 cc.
    Glycerin .... 25 cc.

Allow to settle and decant the clear liquid for use as needed.

With the introduction of dioxan into microtechnique, several workers independently developed the idea of using this reagent in killing and preserving fluids. McWhorter and Weier (1936), for example, devised the following formula:

    Dioxan .... 50 cc.
    Formalin .... 6 cc.
    Acetic acid .... 5 cc.
    Water .... 50 cc.

This solution preserves unicellular and filamentous algae, fungi, and other delicate subjects. Temporary mounts for microscopic study can be made on a slide in a drop of this fluid, or permanent slides can be made.

Temporary and Semipermanent Slides

The foregoing outline of methods of preserving bulk materials leads to a consideration of methods of preparing mounts for microscopic study. The simplest method obviously consists of mounting a small quantity of the material in a drop of water. However, water mounts dry out during prolonged study, and it is better to mount the material in 10% glycerin. As the water evaporates, introduce more glycerin solution under the cover glass at intervals, until no further evaporation takes place. Such preparations can be kept almost indefinitely if stored flat and handled with care.

The preservative and swelling action of lactic acid and phenol (carbolic acid) is utilized in an important class of formulas. More or less durable slides of algae, fungi, fern prothalli, sections, and other small objects can be made by mounting in one of these lactophenol solutions, with or without added dye. The following selected formulas are taken from Maneval's valuable compilation of these methods.

Aman's Lactophenol

Phenol (melted) .... 20 cc.
Lactic acid .... 20 cc.
Glycerin .... 40 cc.
Water .... 20 cc.

Phenol-Glycerin

Phenol (melted) .... 20 cc.
Glycerin .... 40 cc.
Water .... 40 cc.

If a staining effect is desired, add a 1% aqueous solution of either cotton blue, aniline blue, or acid fuchsin as follows:

Lactophenol .... 100 cc.
Glacial acetic acid .... 0 to 20 cc.
Dye solution .... 1 to 5 cc.

The optimum concentration of glacial acetic acid is that which produces no collapse or bursting of cells or filaments. Try the above formula and dilute with lactophenol until the best proportions are established.

Glycerin-jelly preparations are a further advance toward permanent slides and are preferred to glycerin mounts if the slides must withstand considerable use. The mounting medium is made as follows:

Gelatin .... 5 g.
Water .... 30 cc.
Glycerin .... 35 cc.
Phenol (dissolved in 10 drops water) .... 0.5 g.

Dissolve the gelatin in the water at 35°C.; then add the other ingredients. Filter while warm through fine silk or coarse filter paper. This mounting medium keeps well. Materials to be mounted in glycerin jelly are first stained (if necessary), then dehydrated by the glycerin evaporation process. For filamentous algae and fungi the most satisfactory stains are the self-mordanting hematoxylins and iron hematoxylin. Staining trials can be made with small quantities of the plant material until a satisfactory schedule is worked out. Then stain a large batch, and dehydrate by the glycerin method. To make a slide from the dehydrated material, place a piece of glycerin jelly about as large as a match head on a clean, dry slide, and warm until melted. Remove a quantity of the plant material from the pure glycerin, draw off excess glycerin with filter paper, and put the plant material into the warmed jelly. Lower a cover glass carefully over the material. If the material is not excessively fragile, a lead weight on the cover glass will squeeze out excess jelly and make a thinner mount. When the jelly is cool, clean off any excess around the cover glass and seal around the edge of the cover glass with a quick-drying lacquer such as Duco. Sealed preparations will keep for several years, but it is well to remember that the mounting medium is soft. Such preparations are not desirable for use by elementary students.

Permanent Slides of Whole Mounts

Permanent stained slides in a hard, durable mounting medium are much more satisfactory than soft, easily damaged, temporary or semipermanent slides. Modern methods make possible the rapid, quantity production of permanent slides almost as quickly and easily as the making of old-fashioned semipermanent slides. Filamentous algae and delicate objects may be killed in any of the formulas described in the foregoing pages. Wash out the killing or storage fluid with water and apply an appropriate stain. Any of the self-mordanting hematoxylins will give excellent results. It is best to overstain strongly, leaving the material in the stain for ½ to 1 hr. Wash in distilled water until the rinse water is no longer tinted. Destain in 1/10% HCl. Cover the material with the acid in a shallow dish and agitate gently. After 1 to 2 min. drain, rinse in tap water, and examine with a microscope. Repeat the treatment in acid until only the nuclei and pyrenoids remain blue-black, then wash thoroughly in tap water.

Iron hematoxylin will give the most critical staining of nuclear structures and pyrenoids. Mordant algae for 1 to 2 hr. Rinse quickly but thoroughly in distilled water and stain for 2 to 8 hr. Destain by immersing in the destaining agent for 1 to 2 min., rinse in *distilled* water, examine with the microscope, and repeat the brief immersion in alum until the nuclei and pyrenoids are sharply differentiated. After the last rinsing in distilled water, wash thoroughly in tap water. Dehydration and mounting may be accomplished by one of the several methods outlined below.

THE VENETIAN TURPENTINE METHOD

The Venetian turpentine method yields excellent preparations, but this method is likely to be supplanted by modern methods using a variety of synthetic organic solvents and synthetic resins. Therefore, only a brief outline of the method will be given, and the interested reader is referred to Chamberlain for details (1932). Kill and stain the material, and dehydrate by the glycerin evaporation method. Rinse out the glycerin in 95% alcohol, then complete the dehydration in at least three changes of absolute alcohol. Transfer to 10% Venetian turpentine in absolute alcohol. Eliminate the alcohol by evaporation. When the material is in thickened turpentine, mount the desired amount of the material in a drop of that medium. Dry the slides in a horizontal position. Preparations made by this method are durable and the stains are permanent.

THE BUTYL ALCOHOL-RESIN METHOD

The term resin is used here in the broad sense to include Canada balsam as well as the increasing number of synthetic resins. Test the solubility of the currently available mounting resins in *anhydrous* normal and tertiary butyl alcohol and in dioxan, and use the solvent-resin combination that has the highest transparency and least color. The following tertiary butyl alcohol – Canada balsam procedure serves as an example of the series of operations in a whole-mount method. Stain and wash as before. Transfer a small quantity of the material into a watch glass of 50% alcohol, and observe with the microscope. If there is much distortion, try 20% alcohol on another batch. Well-hardened material can withstand 50%. When the proper starting point for dehydration is established, carry the material in steps of 20 to approximately 70% alcohol. Add to the 70% a few drops of stock solution of counterstain — eosin Y, erythrosin B, or fast green, saturated solution in absolute alcohol or in methyl Cellosolve. Leave in the counterstain until slightly overstained. This may require 4 to 12 hr. Rinse in 70% alcohol, and transfer through the following series at ½- to 1-hr. intervals:

3 parts alcohol to 1 part *TBA* (anhydrous tertiary butyl alcohol)
2 parts alcohol to 2 parts *TBA*
1 part alcohol to 3 parts *TBA*
Pure *TBA;* change twice at 15-min. intervals.

Transfer to a large volume of 5% solution of balsam or synthetic resin in *TBA* in a short wide-mouthed bottle. Allow the *TBA* to evaporate slowly at a temperature of about 35°C. When the balsam is slightly more fluid than that used for covering sections, mount the material. Remove a suitable quantity of the plant material with its enveloping balsam, place on a dry, clean slide and lower a dry cover glass carefully so as not to produce bubbles or to push the plants too close to the edges. Dry the finished slides in a horizontal position. (Johansen, 1940.)

THE DIOXAN-BALSAM METHOD

This is the most promising of the newer methods of making whole mounts. The method was worked out in almost identical form independently by McWhorter and Weier (1936), Johansen (1937), and the present writer. It is probable that numerous other workers had developed similar schedules.

Stain filamentous or other delicate materials as in the preceding methods. Pass through a series of aqueous solutions of dioxan, containing the following percentages of dioxan: 20, 40, 60, 80, 90, then three or more changes of *anhydrous, chemically pure* dioxan. The interval in each should be 1 to 2 hr. Examine a few filaments under a microscope, mounted in the last fluid. If the material is in good condition, transfer to a 10% solution of balsam or resin in dioxan. Use a wide-mouthed bottle and gauge the volume of the liquid so that the material does not become exposed as the dioxan evaporates. Place the uncovered container into an oven or a dust-free place at a temperature of approximately 35°C. The dioxan evaporates in 2 to 8 hr., leaving the material in thick resin in which it is mounted. Extremely fragile materials, such as *Volvox* or *Vaucheria,* should be started in 5% resin and evaporated very slowly by keeping the container loosely covered.

This process is so rapid that it is worth the time to carry small trial lots of the material through the process at different speeds. The condition of the material can be examined at various points in the process. When the optimum or shortest safe schedule is found, the main batch of material can be carried through.

ACETOCARMINE (OR ACETO-ORCEIN) SMEARS

The acetocarmine method has become so commonplace that it may well be included in an elementary manual. This rapid method combines killing, fixing, and staining. The freshly-made preparations can be used in nonpermanent form for counting chromosomes, determining their association, and studying intimate details of structure. The slides can be made permanent if desired.

The stain is prepared by dissolving 1 g. carmine in 100 cc. of boiling 45% acetic acid (or propionic acid). Cool and decant. Add 2 drops of a saturated aqueous solution of ferric acetate, and allow to stand for 12 hr. Filter and store the main stock in a refrigerator, keeping a small quantity in a dropper bottle in the laboratory for immediate use. Some workers omit the iron salt and incorporate the correct amount of iron from the reaction between steel dissecting needles and the acetic acid in the dye. This requires considerable experience. Excessive iron prevents clear differentiation and may produce a dark granular precipitate. This can be minimized by cleaning the steel needles in 45% acetic acid periodically during dissection. If iron is added to the dye formula, use nickel or chromium plated needles or pyrex needles drawn to an abrupt, fine point.

The simplest method of using acetocarmine consists of macerating or smearing a fresh anther in a drop of the stain. Small anthers may be dropped into the dye entire, then dissected under a binocular, discarding pieces of anther wall and leaving only the masses of sporocytes. Large anthers should be sliced into the thinnest possible slices on a sheet of wet blotting paper, the fragments transferred to a drop of dye and macerated as above.

Lower a cover glass over the drop of dye containing the sporocytes, and press or tap gently. A careful sliding motion on the cover glass sometimes aids in smearing the cells into a thin film. Pass the slide quickly over an alcohol lamp flame several times. The amount of heating must be determined by trial, but do not heat to the boiling point. Drain off excess stain, and seal the edges of the cover glass with paraffin. Examine the slides with a microscope, and store satisfactory ones in a refrigerator. The color improves in a few days, reaches maximum intensity, then begins to deteriorate.

Fresh anthers are not available at all times, and it is often necessary to make extensive collections in the field for subsequent study.

Kill entire anthers, or grass spikelets, or even large portions of an inflorescence in Farmer's or Carnoy's fluid. Maize cytologists prefer an alcohol-acid ratio of 3:1, but ratios of 1:1, 2:1 etc. may be tested until the optimum fixation is obtained for a given species or stage. Plant materials can be stored in these fluids for months at 0°C., depending on specific response. Most workers transfer the tissues after 1 or 2 days of fixation to 70% alcohol, in which tissues can be stored for many months in a refrigerator. Good preservation for as long as a year can be obtained by storing in acetic-alcohol in a freezer.

Root tip smears are being used extensively for chromosome counts and for the study of the morphology of metaphase chromosomes. The principal problem is to obtain separation and flattening of cells. Some large root tips, such as onion, may be squashed flat easily, whereas the small root tips of sweet clover resist crushing.

Fix the root tips in Farmer's or Carnoy's fluid for at least one day. Smear directly from the fixing fluid, or transfer to 70% alcohol for prolonged storage. The responses of a species to the fixing formula, the fixing interval and the permissible storage period in fixing fluid or 70% alcohol, must be determined by experimentation.

The piece of root tip that is fixed is usually at least three times as long as the most actively meristematic region. Cut off this active end into a drop of acetocarmine, and macerate into linear strips with needles. Some materials are benefited by some heating at this stage. Cover and heat carefully. Press or smear the cover until the cells are sufficiently separated and flattened.

The separation of cells can be improved by hydrolyzing the middle lamella with 5 to 10% HCl. The acid may be in water, or 70% alcohol, or in Farmer's fluid. After immersion in the acid for 5 to 30 min., return the roots to fixative, which should be changed at least once. Enzymes also have been suggested for digesting the middle lamella.

Clumping of chromosomes can be prevented, and some morphological details accentuated by pre-treatment of the living root tips. Soak the tips in a saturated aqueous solution of paradichlorobenzene for 1 to 4 hours and fix in the preferred formula. Methyl alcohol, in concentrations of 1% to 3% in water has also been used for pre-treatment.

Several readily available plants, especially *Rhoeo* and *Tradescantia,* yield excellent sporocyte smears that show the coiled chromonemata (Sax and Humphrey). Smear sporocytes on a clean slide and flood with 1% ammonium hydroxide in 30% alcohol. Make trials at intervals of 10–30 seconds. Drain and cover with acetocarmine. The time and temperature of heating at this stage must be determined by trial. The slide can be studied as a temporary slide, or made permanent. Variations of the acetocarmine process can be found in Smith's comprehensive review.

THE FEULGEN METHOD

The Feulgen smear or squash technique has become firmly established in cytological research. The process and allied methods are under intensive study in many laboratories; therefore, only a few citations are given here. Lillie described an error-proof method of making the Schiff reagent for the Feulgen stain (1951). Dissolve 1 g. basic fuchsin and 1.9 g. $Na_2S_2O_5$ in 100 cc. of 0.15 N HCl. Shake on a mechanical shaker for two hours. Add 0.5 g. fresh activated charcoal, shake for two minutes, and filter. If the reagent becomes pink, add charcoal, shake, and filter. Store the pale yellow reagent in a refrigerator.

The reagent can be used with sectioned material (Lillie, 1954). Many dyes other than fuchsin can be used to make Schiff-type reagents (Kasten, 1950).

To make squash preparations, fix tissues in Farmer's fluid. Hydrolize in N HCl for 8–10 min. at 60°C. The time and temperature are critical. Drain and cover with Feulgen reagent. After 1 hr. at room temperature, transfer a piece to 45% acetic acid and squash with a glass rod. Cover and seal, or make permanent.

Smear preparations can be made permanent. The process consists of dehydrating the cells and mounting in a resin. If the cells are smeared on perfectly clean glass, they will adhere during subsequent treatment. Slip the cover glass by immersing the slide in equal volumes of acetic acid and alcohol, inverted in a Petri dish, with one end of the slide held up on a glass rod. Pass the slide and its cover glass through the following fluids at 2–5 min. intervals:

(1) equal volumes of ethyl and tertiary butyl alcohol
(2) tertiary butyl alcohol, 3 changes

Place the slide with the smear upward on blotting paper, place a drop of thin balsam or resin on the smear, and lower the cover glass carefully. Place a lead weight on the cover glass. A similar acetic acid-ethyl alcohol-xylene series can be used (McClintock) but more intermediate gradations must be used to prevent collapse of the sporocytes.

A quick-freeze method of making slides permanent, using dry ice (Conger and Fairchild), paved the way for an even better method using liquid carbon dioxide (Bowen).

MACERATION

Whole mounts or sections of tissue frequently do not convey an adequate three-dimensional impression of cell structure. A valuable and much-neglected type of preparation is made by isolating complete individual cells from a mass of tissue. This is accomplished by chemical and mechanical separation of cells. Divide the material into slivers thinner than a toothpick. If the material is dry, boil in water containing a wetting agent, aspirating if necessary. Treat with one of the following macerating processes:

Schultze's Method.—Cover the material with concentrated nitric acid.
Add a few crystals of potassium chlorate.
Heat gently on a sand bath, in a closed hood, until the material is bleached white.
Wash thoroughly, and shake with glass beads until the material disintegrates.
Increase or decrease the duration of heating until entire unbroken cells can be isolated.

Jeffrey's Method.—The macerating fluid consists of equal volumes of 10% chromic acid and 10% nitric acid.
Treat for 1 to 2 days at 30 to 40°C.
Wash and shake with glass beads.

Harlow's Method.—Treat the subdivided and boiled material in chlorine water for 2 hr.
Wash in water.
Boil in 3% sodium sulphite for 15 min.
Wash and macerate.
Repeat chlorination and the sulphite bath if necessary.

Following any of these maceration treatments, wash the pulp thoroughly by decantation. A centrifuge is an aid in washing. Unstained material may be mounted in water or glycerin for study. The cells may be lightly stained in aqueous safranin, washed, and mounted. The mounting media described on pages 102–5 may be used to make semipermanent or permanent slides.

11. Criteria of Successful Processing

We have carried to completion the processing of some plant materials and have the finished slides before us. Our notebook contains a record of the entire process, from collecting the living plants to the completion of staining. We can now examine the slides critically and consider the criteria by which we may judge the success or failure of the processing. The severity of our scrutiny depends on the objectives of the study for which the preparations were made. We may be primarily interested in the distribution of tissue systems or the position of organs, paying little attention to the protoplasm. Rapid and convenient methods that preserve the desired structures with adequate accuracy need not be regarded as slovenly. Another study may require the preservation of the constituents of the protoplasm in their normal structure and position. The use of more elaborate, time-consuming processes is then justified.

It is axiomatic that slides for elementary students must be much more perfect than for advanced workers. Beginners waste much time puzzling over imperfections. They will draw faithfully a break produced in microtoming, the gap between the cell wall and the collapsed plasma membrane, or a speck of debris in the tissues.

The pathologist should be especially critical. Control slides of normal tissues must show cellular details with almost diagrammatic perfection in order to furnish a basis for interpreting pathological conditions. Studies in physiology, cytology, and experimental morphology also demand careful control of processing and critical examination of slides on the basis of some useful criteria.

The following illustrations were selected from well-known and widely used subjects. Leaves of different species exhibit different reactions to killing fluids. The quality of the fixation can be determined easily by the condition of the palisade cells. The outlines of the cells should not be wrinkled, and the chloroplasts should line the walls. Figure 11.1 *a* shows the excellent preservation of cellular and tissue details in soybean leaf, a highly spongy and therefore difficult subject, processed as outlined in the legend.

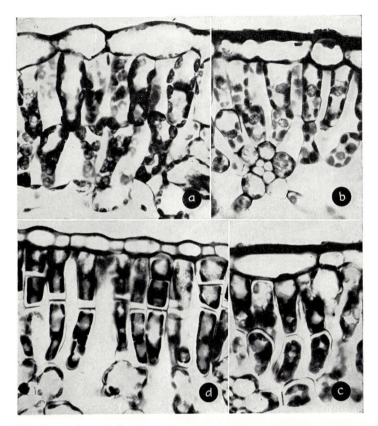

Fig. 11.1—Comparison of fixation in leaves: *a,* soybean, Craf III, acetone-tertiary butyl alcohol; *b,* cherry, turgid condition, Craf II, alcohol–xylene; *c,* cherry, plasmolyzed, *FAA,* alcohol–xylene; *d,* red ash, plasmolyzed, *FAA,* alcohol–xylene.

A striking illustration of the effects of different killing fluids is afforded by experiments with young leaves of cherry. Figure 11.1 *c* shows the severe plasmolysis obtained by killing in *FAA*. Figure 11.1 *b* shows the turgid, normal condition of the palisade and spongy parenchyma, following killing in a Nawaschin type formula, Craf II. After killing, both lots were processed simultaneously by identical methods, an alcohol-xylene series and careful infiltration in paraffin.

Stems are subject to the same defects as the leaves described above. Study Fig. 11.2 *a,* a cross section of alfalfa stem. Note that the delicate hypodermal chlorenchyma is intact, the cells have smooth, rounded outlines, and there is no marked cleavage between cells. Within each chlorenchyma cell the plasma membrane and the chloro-

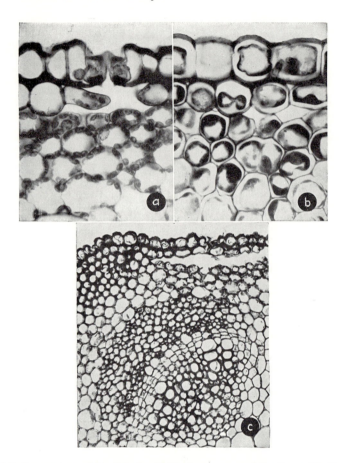

Fig. 11.2—Comparison of fixation in alfalfa stems: *a,* good fixation with Craf 0.20–1.0–10.0, dioxan process; *b,* plasmolyzed cells in cortical region, Craf 0.30–1.0–10.0, acetone-*n*-butyl alcohol process; *c,* good fixation, but with peeling epidermis, process as in *a.*

plasts are appressed against the cell wall, indicating absence of plasmolysis. The physical appearance of these cells may be regarded as approximating the structure of living cells. Observe further that the epidermis is not peeled away from the inner tissues. These features are indicative of successful processing.

In comparison with the above case, examine Fig. 11.2 *b,* also a cross section of alfalfa stem, killed in *FAA* and carried through an alcohol-xylene series. There is marked plasmolysis; the massed chloroplasts and cytoplasm absorb the stain and cannot be destained differentially, producing harsh staining effects. Note especially the collapsed plasma membranes in the epidermis and chlorenchyma. This embedded material was brittle and only the pieces from the upper three internodes could be cut without serious tearing.

Figure 11.2 *c* is from a third batch of alfalfa stem which was processed exactly like the material from which Fig. 11.2 *a* was obtained. The structure of the deeper cortical cells is well preserved, but the epidermis shows extensive peeling away from the chlorenchyma. The condition of the protoplasts indicates that the killing fluid and subsequent processing were not at fault. It is probable that the peeling of the epidermis was caused by compression of the extremely soft hypodermal tissues when the fresh stem was cut into pieces for killing.

Roots are somewhat more difficult to judge than organs containing chloroplasts. Root cells that contain leucoplasts should be examined for the position of these plastids. If the plastids are inconspicuous, the condition of the thin plasma membrane will indicate whether plasmolysis has occurred.

The critic must be much more lenient in the examination of sections of wood. Boiling the wood in water, desilicification with hydrofluoric acid, and infiltration under pressure for long periods are not conducive to preservation of protoplasmic or even cell-wall details. Mechanical tearing can be easily recognized. In the nonliving elements (tracheae and tracheids) the concentric layers of the thick wall should not be separated. The innermost layer is often found to be completely separated in poor preparations. Parenchymatous elements, such as wood rays, diffuse wood parenchyma, and epithelial cells of resin canals, are subject to plasmolysis. Perfect fixation of these parenchymatous cells should not be expected.

The embryo sac of lily is a difficult subject that tests the effectiveness of a method and the skill of the worker. The young megasporocyte of lily is comparatively easy to preserve in good condition. Such fluids as chrome-acetic and Bouin's, followed by the traditional alcohol-xylene series, yield excellent preparations and a good ratio of well-preserved sporocytes. The sporocyte and integument initials in Fig. 11.3 are preserved very well indeed for routine class material.

With continued enlargement of the sporocyte the cytoplasm probably becomes highly fluid, the integuments elongate as thin sheets of tissue, and these structures become increasingly subject to damage. Look for evidence of plasmolysis of the sporocyte and wrinkling of the rims of the integuments. In Fig. 11.4 *a* the finely granular cytoplasm is obviously not shrunken, and the rounded rims of the integu-

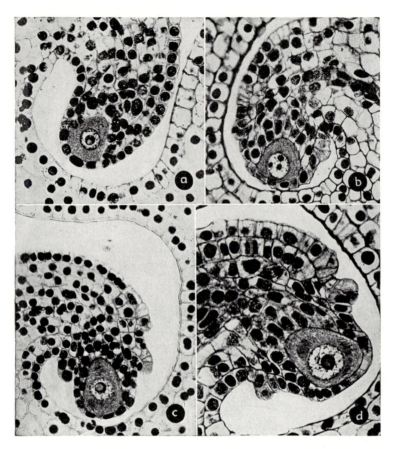

FIG. 11.3—Comparison of fixation in young ovules of *Lilium*: *a, L. speciosum,* chrome-acetic (0.5–0.5) 48 hr., alcohol-xylene; *b, L. tigrinum,* Allen-Bouin III, 4 days, three-grade dioxan; *c, L. speciosum,* chrome-acetic (0.5–0.5) 48 hr., alcohol-xylene; *d, L. tigrinum,* Allen-Bouin III, 6 months, acetone-tertiary butyl alcohol.

ments show very little distortion. Many readers will recognize an old acquaintance in Fig. 11.4 *b,* a typical Bouin, alcohol-xylene image that is frequently encountered in lily ovule. Although some batches prepared by this method have some good sporocytes, severe plasmolysis occurs frequently. However, nuclear fixation is usually excellent.

Figure 11.4 *c* shows very nearly perfect preservation of the sporocyte. The cytoplasm in Fig. 11.4 *d* is somewhat coarsely granular, but the chromosomes at the beginning of disjunction in the first division of meiosis are well fixed. Note that each chromosome is long and U-shaped, instead of a compact lump as with inferior fixation.

It is probable that undue blame has been placed on the killing fluid for the distortion of protoplasts of these large sporocytes. Recent developments in dehydrating and infiltrating methods indicate that these processes contribute much to the quality of the image. Rapid dehydration is probably one cause of distortion. The slow dehydration obtained by the glycerin evaporation method yields a very high ratio of well-preserved ovules. The mild dehydrating action of such paraffin solvents as the butyl alcohols and dioxan minimizes distortion of large sporocytes. It is probable that a properly balanced killing fluid,

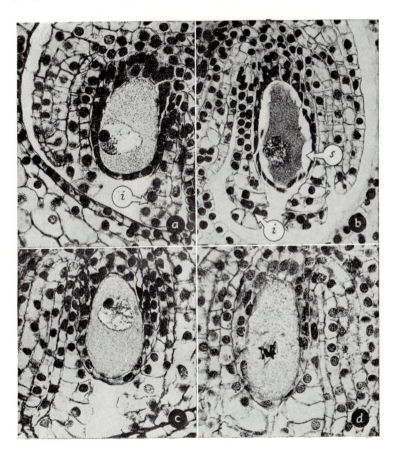

Fig. 11.4—Comparison of fixation in large megasporocytes of *Lilium*: *a, L. umbellatum,* Allen-Bouin III, 12 days, glycerin dehydration to alcohol-xylene; *b, L. speciosum,* Bouin, alcohol-xylene; *c, L. tigrinum,* Allen-Bouin II, 6 months, acetone-xylene; *d, L. tigrinum,* first division of meiosis, Allen-Bouin II, 6 months, acetone-xylene.

followed by a mild and slow dehydrating method, and progressive infiltration, yields the most satisfactory ratio of good ovules.

The foregoing discussion of the criteria of the quality of processed materials was illustrated by rather extreme cases of unsuccessful processing, contrasted with some decidedly choice materials. In actual practice it is not necessary to be so critical. A preparation having some plasmolysis may nevertheless be presentable and usable for the study of organogeny and tissue systems. At the risk of tiresome repetition it will be stated again that the choice of the subject, the technique used, and the quality of the finished product should suit the need.

Part II

Specific Methods

12. Introduction

The presentation of detailed directions for specific subjects must be prefaced by certain reservations and precautions. Reactions of plants vary with their age, degree of differentiation, degree of turgor, and perhaps many intangible physiological factors. Each step in the processing involves a time factor, the duration of that particular treatment. The numerous successive operations, each with a variable time factor, offer innumerable combinations of treatment. This is complicated by variations in the purity of reagents, fluctuations of room temperature, failure of oven thermostats, and just plain blunders by the worker. The possibilities of influencing the finished product are obvious. Therefore the author of a research paper or a manual is reluctant to present a set of written directions with any assurance of success to his readers. However, a study of the general principles of collecting, killing, and processing, as outlined in Chaps. 1, 2, and 3, will enable the reader to use the following directions with reasonable assurance of success.

This section of the manual should be regarded as closely linked with Part I. The general methods of collecting and processing described in Part I will now be supplemented by specific recommendations as to suitable plants for illustrating various topics and the techniques of processing these plants. The reader should refer back frequently to pertinent portions of Part I.

Plants selected for study not only should show the desired structural features to best advantage but also should have the virtue of familiarity to the student and availability. Why use *Vanilla,* a rare orchid, to illustrate the anatomy of the monocot root when the common garden asparagus yields instructive slides? Some of the slides used for teaching should be of species other than those illustrated in the official textbook. This demands more critical study of the slides by the students and minimizes the copying of text figures. The plants recommended here are available in most parts of the country or can be grown in the greenhouse or even in a window box. The local florist shop and commercial green houses are fruitful sources of materials, especially exotics. Algae, fungi, and bryophytes can be found in abundance when one has learned where to look. Such local foraging and field trips afford a wealth of material.

The sequence in which specific recommendations are arranged in this manual takes cognizance of the customary arrangement of topics in textbooks of botany. Laboratory courses in general botany, anatomy, and histology are oriented around topics and fundamental problems that cut across taxonomic lines. The leaf, for example, is studied as a functional and structural unit, a food-making organ. A comprehensive study of the leaf from this point of view necessitates a comparison of leaves of a wide range of vascular plants and perhaps even of mosses. The stem and root are likewise studied as organs having structural diversity and functional modifications but nevertheless having some fundamental pattern. In addition to that elusive entity, the *typical* organ, it is essential to examine variations and modifications of the basic pattern. A comprehensive study of vascular anatomy thus embraces vascular plants from *Lycopodium* to *Orchis*.

From the standpoint of technique each organ presents its characteristic problems. For example, broad leaves of plants in widely separated taxonomic groups have in common such problems as collecting, subdivision in sampling, and orientation in sectioning. If we were to consider in its entirety some one species, like an oak, we would find that its root tips, embryo sac, and old stem present very different problems of technique.

These considerations have led to the arrangement of Part II, in which categories of organs as well as taxonomic position are used as major chapter topics.

13. Vegetative Organs of Vascular Plants

The vegetative organs of the vascular plants are the leaf, the stem, and the root. These organs can be studied either from the standpoint of developmental morphology and histogenesis, or they may be studied from the comparative viewpoint by a comparison of the mature organ in its diverse forms. A combination of the two viewpoints has much merit, and the following presentation of materials and methods affords suitable material for such studies.

Meristems

This section is limited to the apical meristems or growing points of stems and roots and the associated organ primordia. Lateral meristems are more properly discussed in connection with secondary growth of older stems and roots. The study of the activities of meristems is in part a study of mitosis. Some prepared slides of meristems are intended to show critical details of mitosis; however, some slides also are prepared to show tissue systems and organ primordia. For either type of slide, meristematic tissues are processed by the most critical methods that time and facilities permit.

THE ROOT TIP

Growing points of roots are obtainable from seedlings sprouted on blotting paper, from sprouted bulbs, from older plants in pots, or from plants dug up in the field. Regardless of the source or length of a root, the meristematic region is confined to the terminal 1 to 2 mm. (Fig. 13.1). Penetration of reagents occurs over the entire surface. For elementary work, adequate cellular detail is obtained with the entire root tip. The pieces are thus large enough to be handled easily. Longitudinal sections show the relationship of the root cap, the meristem, and the older tissues, and slides can be made by quantity-production methods. For more critical studies the root tip should be cut into the smallest possible pieces or prepared by one of the modern smear methods. The best of these methods have been admirably assembled by Johansen (1940) and Smith (1947). There is a daily periodicity in the rate of mitotic activity, a periodicity that is characteristic for each species. In order to have many mitotic figures on the slides, root tips should be collected at periods when many cells are dividing.

Bulbs of *Allium cepa, Hyacinthus, Crocus, Tulipa,* and to a lesser extent of *Lilium* are good sources of root-tip material. The following suggestions, based on onion, will furnish the basis for other related subjects. The simplest method of sprouting the bulbs is in individual containers of water, using a tall narrow bottle, jar, or drinking glass of such diameter that the lower third of the bulb is submerged. Change the water twice a day. Bulbs sprout well in moist, steam-sterilized sphagnum. The peaks of mitotic activity for onion are from 1:00 to 2:00 P.M. and 11:00 P.M. to midnight.

Onion root tips are preserved satisfactorily in fluids of the Na-waschin type; one of the most consistently satisfactory is III. The comparatively expensive butyl alcohol and dioxan methods give good results, but a closely graded acetone-xylene series and careful infil-tration yield excellent preparations for general class use. Bouin's solution yields excellent mitotic figures, especially prophases and telophases. Occasional lots killed with Bouin are extremely poor. Fluids containing osmic acid are favored by some workers for critical cytological work, but the use of this expensive reagent is not justified for routine work. The formula must be adapted to the plant being studied, and the reader who wishes to use such fluids must consult the research literature for details.

Safranin-fast green and the triple stain give the most complete picture of the entire cell, differentiating the cell wall, the texture of the cytoplasm, the achromatic figure, and nuclear structure. Iron hematoxylin stains chromatin an opaque, contrasty blue-black against a gray cytoplasm. Gentian violet-iodine gives a brilliant blue-black translucent chromatin on a perfectly colorless, almost invisible back-ground. Select the stain combination that gives the desired effect.

Hyacinthus and *Narcissus* also are suitable sources of root-tip preparations. The methods are essentially as given for the onion. *Gladiolus* has very small chromosomes, and preparations are of value mainly to illustrate comparative chromosome sizes. The possibilities of the numerous kinds of bulbs, corms, and rhizomes available in field and garden have been by no means fully explored.

Root tips of corn are obtainable by sprouting kernels in a moist chamber. When seminal roots and some lateral roots have developed, cut off the meristematic tips. Root tips also may be obtained from pot-bound plants and the plants can be repotted without apparent damage. Mitotic activity is usually rapid during the early forenoon. Maize cytologists favor a formula that is practically identical with Craf III. Choose a stain by the criteria discussed in connection with the onion.

Vicia faba, the horse bean, has 12 large chromosomes, 2 of them about twice as large as the others. Obtain root tips by sprouting seeds in a moist chamber or from plants grown in pots of sphagnum. Kill in Nawaschin or in Craf II, and stain as with onion. The radicles of many other legumes also are easy to obtain and to process. Sections may be stained for either histological or cytological details (Fig. 13.2), or a good compromise may be obtained with safranin-fast green.

The common trailing *Zebrina* grown in greenhouses has large chromosomes. Obtain root tips from cuttings rooted in sand. The periodicity is an uncertain factor, and the worker must chance obtaining abundant mitotic figures. Bouin's solution and Craf II usually give acceptable results.

APICAL MERISTEMS OF THE STEM

The origin of the tissues of the stem and of lateral organs on the stem is revealed by a study of the meristematic tip or apex of a stem. This growing apex may be found at the tip of a growing axis, or in a dormant terminal or axillary bud. One of the easiest subjects to handle is the shoot from the sprouting kernel of corn. Sorghum, oats, and other small grains also may be used, but the coleoptile is small and not so easy to handle as that of corn. The growing apex of these Gramineae is a broad dome, from which the leaf primordia arise as lateral protuberances. Successive leaf primordia are laid down during this period of rapid growth and may be seen in graded order in a good median longitudinal section. Transverse sections show the lateral extension of the meristematic margins of the leaf primordia. The oat sprout develops axillary buds earlier than do the other suggested grasses.

To obtain growing apices of stems of *Zea,* germinate the corn in sphagnum or sand. When the coleoptile is approximately 5 cm. long, cut out a section 5 mm. long at the coleoptile node (Fig. 13.1 *A*). This region, which contains the growing apex, can be located easily by holding the shoot before a bright light — the region of the coleoptile node and compact growing apex area is dark. Older seedlings show more advanced axillary buds than the young sprouts. One

week to 10 days after emergence, the seedling has leafy axillary buds
(Fig. 13.3 *a*). The tassel primordia become evident 25 to 30 days
after emergence (Fig. 13.3 *b*). Instructive vegetative and floral
apices can be obtained from the sprouts that arise from sods or
clumps of the larger perennial grasses, such as brome grass or orchard
grass. Inflorescences are initiated during April (Fig. 13.3 *c*), and
vegetative tips are obtainable from the new sprouts that emerge in
midsummer or later. Excellent preservation of these gramineous
growing apices can be obtained with Craf I. The air must be completely evacuated from the tightly overlapping leaves encircling the
stem tip. Improper infiltration results in the collapse of the meristematic tissues and breaking of the tip and leaf primordia during
sectioning. Cut sections of maize 10 to 12 μ thick and Avena 6 to 8 μ.
Stain in tannic acid-ferric chloride, in hemalum-safranin, or in
safranin-fast green.

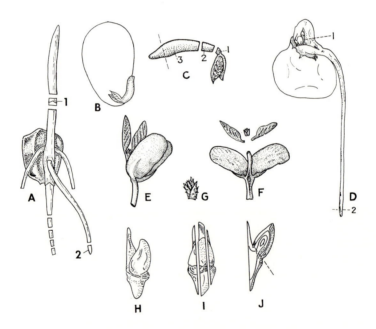

Fɪɢ. 13.1—Methods of obtaining apical meristems. *A*, seedling of corn, growing
apex of stem is at coleoptile node 1, root tip removed from seminal root 2; *B*, half
of kidney bean seed with embryo in place; *C*, parts of embryo, 1 contains stem tip,
2 is discarded, 3 is used for root tip; *D*, sprouting pea, epicotyl cut off at 1 is used
for stem tip, root tip cut off at 2; *E* and *F*, sprouting soybean dissected to obtain
terminal bud *G*; *H–J*, bud of basswood removed from twig and divided for killing.

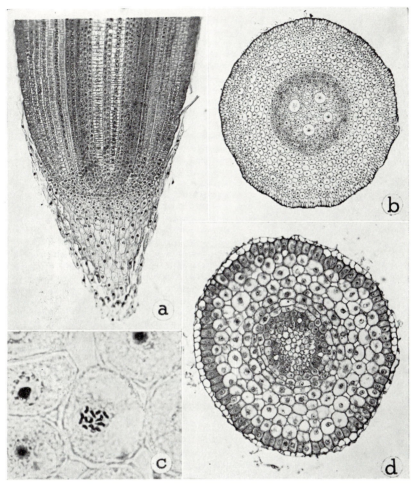

Fɪɢ. 13.2—Root tip meristems and histogens. *a, b, Zea mays; c, Melilotus alba; d,
Lotus tenuis,* chromosome complement, iron-hematoxylin.

A good dicotyledonous stem growing apex can be obtained from
young seedlings of Lima beans, kidney beans, soybeans, flax and peas,
sprouted in sphagnum or sand. Lima and kidney beans have a large
epicotyl of simple structure at the postdormant stage, after the seed
coat has been ruptured but the plumule has not yet emerged. Peel off
the seed coat, separate the cotyledons, and remove the epicotyl and
radicle (Fig. 13.1 *B, C*). Good fixation can be obtained with Craf II
followed by the acetone-xylene series. Cut the paraffin sections at
right angles to the flat, overlapping plumule leaves. Stain as

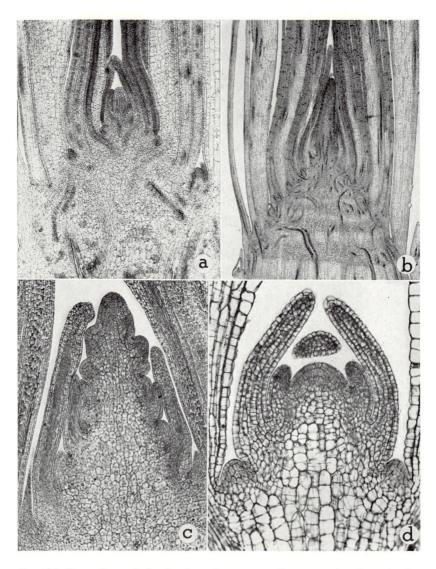

FIG. 13.3—Vegetative and floral apices of stems: *a, Zea,* vegetative, 1 week after emergence; *b, Zea,* tassel primordia, 7 weeks after emergence; *c, Bromus inermis,* inflorescence primordia in April; *d, Linum,* plumular bud of seedling.

recommended for corn seedling. The median section will show a broad apical meristem, two small leaf primordia, and fragments of the folded plumule leaves. The radicle can be used for histological or cytological preparations of the growing point.

Peas show a more advanced condition at a corresponding stage of germination. The pea bud is perfectly glabrous. Sprouts showing axillary bud primordia are obtained when the sprout has emerged from the seed in the form of a loop (Fig. 13.1 *D*). Increasing complexity develops rapidly as the sprout becomes straight.

The epicotyl in sprouting soybean is more advanced in organization than in beans or peas. Extract the soybean bud from the burst seed. For an older stage, permit the epicotyl to elongate until the tips of the plumule leaves just protrude beyond the cotyledons. Remove the cotyledons, pull the plumule leaves apart, and remove the entire bud (Fig. 13.1 *E–G*). Soybean buds are pubescent and must be pumped with an aspirator until they sink in the killing fluid. Large multicellular hairs in the axils of the leaf primordia are easily mistaken for axillary buds by elementary students. The bud is a desirable item for advanced teaching (Fig. 13.4 *b*). The growing apex of the flax seedling is glabrous and very simple in organization (Fig. 13.3 *d*). When the appressed cotyledons of the seedling have begun to diverge, make a transverse cut 1 mm. below the cotyledonary node and another cut 2 mm. above the node. The cotyledons serve as a guide to orientation in the microtome. Section at right angles to the flat sides of the cotyledons.

Axillary buds of *Coleus,* tomato, and other herbaceous plants or the buds from potato eyes also are desirable subjects. Before dropping buds of this type into the killing fluid, it is best to dissect away some of the outer bud scales.

Uniformly good fixation of buds of the above legumes and other recommended herbaceous plants has been obtained with Craf II and the dioxan method, the ethyl-normal butyl series or the dioxan-normal butyl series. Some good results, with occasional unexplained failures, have been obtained with dioxan alone.

Buds of trees and shrubs collected at different seasons show the initiation of leaves and flowers, or the dormant condition. Expanding spring buds of the maples, basswood (*Tilia glabra*), and tulip poplar (*Liriodendron tulipifera*) are recommended. In the shrubs, lilac (*Syringa*), honeysuckle (*Lonicera*), and elderberry (*Sambucus*) are excellent subjects. Remove the buds from the twig as shown in Fig. 13.1 *H–J*. Slice off a longitudinal slice from each side, peel off some

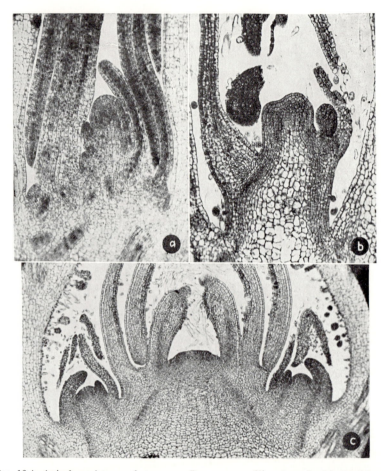

Fig. 13.4—Apical meristems of stems: *a, Zea mays,* seedling 1 week old, Craf I, ace-tone-xylene; *b, Glycine max,* seedling 1 week old, Craf II, three-grade dioxan; *c, Acer saccharinum,* dormant bud from large tree, *FAA,* alcohol-xylene.

of the tougher outer scales under a magnifier, drop into the killing fluid promptly, and pump vigorously. *FAA* penetrates well and gives acceptable fixation (Fig. 13.4 *c*). If more perfect preservation of the protoplast is desired, dissect away most of the larger leaves, and kill the remaining growing apex and small leaf primordia in Craf II. Dehydrate the more brittle buds in butyl alcohol. Use any of the stains recommended for previous growing points.

Fig. 13.5—Stem of *Zea; a,* sector; *b,* single bundle near periphery of stem.

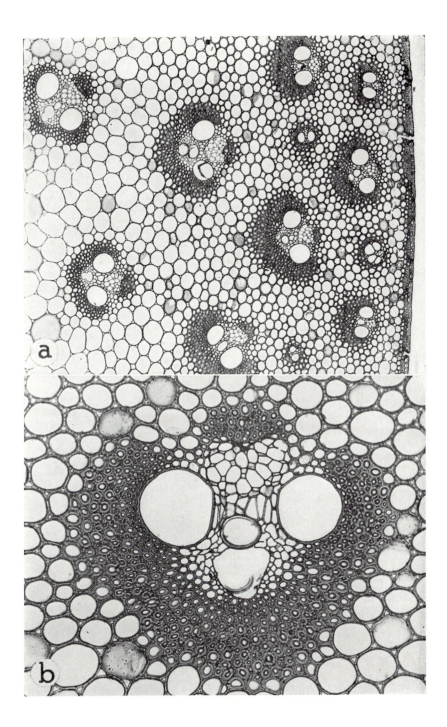

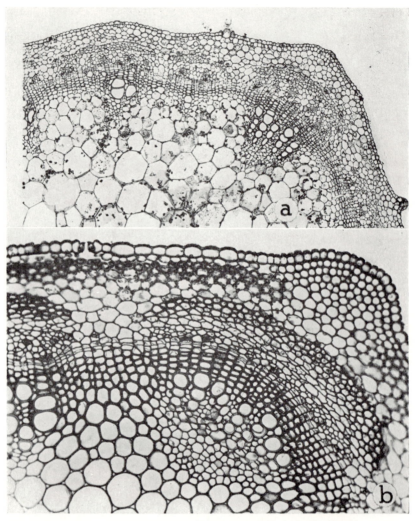

FIG. 13.6—*a*, Stem of soybean, *Glycine max; b*, stem of alfalfa, *Medicago sativa*.

The Stem

The techniques used for the processing of stems range from the foregoing methods used for the delicate meristematic tip to the rather drastic and apparently crude methods necessary to make slides of seasoned lumber. The portion of stem to be selected for sectioning depends on the degree of differentiation that is to be demonstrated.

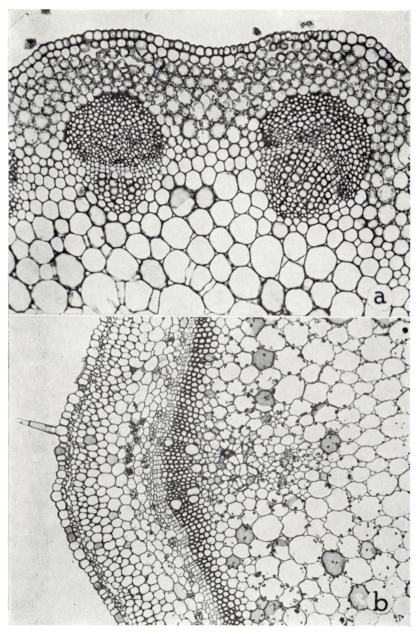

FIG. 13.7—*a*, Stem of alsike clover, *Trifolium hybridum; b,* stem of tomato,
Lycopersicum esculentum.

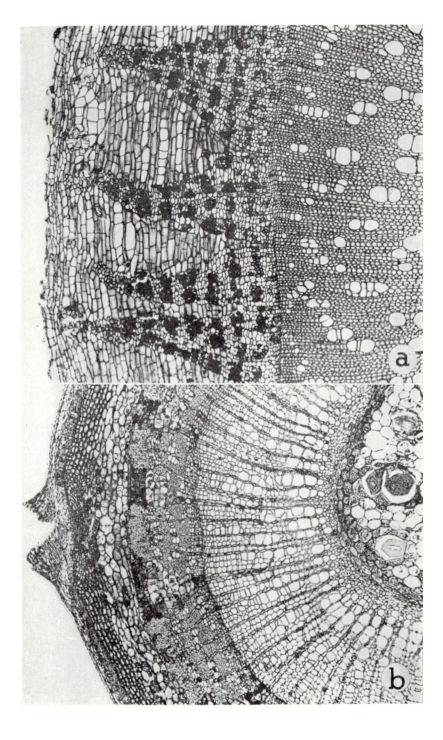

MONOCOTYLEDONOUS STEMS

It is convenient to discuss first the monocotyledonous stem because these stems reach a climax of differentiation in one growing season and do not present the problems raised by the secondary growth of dicotyledonous and gymnosperm stems. Maize may well be used as the standard subject for the grass stem. Complete transverse pieces of seedlings will show the overlapping whorls of leaves encircling the stem. Nodal pieces show the axillary buds, the potential ears. From the older plants use only the internodal pieces of stem, stripping away the leaves. A pot-grown plant will become fairly well lignified and yet be so small that a complete cross section, or at least a quarter sector, can be placed on a slide. However, such plants give an inaccurate picture of the number and structure of the bundles. To show the well-developed and lignified bundle sheath and cortex or rind, use large, field-grown plants at about the time of pollination. Cut the stem into short disks, and divide each disk longitudinally.

Young stems collected before the internodes have become exposed should be killed in a mild fluid like Craf II. Mature stems must be killed in *FAA* and pumped until they sink. The dry, air-filled pith is difficult to infiltrate; it is therefore desirable to exhaust again in the anhydrous stage of dehydration. The use of normal or tertiary butyl alcohol permits paraffin embedding of all but the toughest stems, which must be cut in celloidin (Fig. 13.5). Transverse and longitudinal sections of corn stem take a brilliant safranin-fast green stain. The hemalum-safranin combination is the second choice. Iron hematoxylin-safranin is used only if the middle lamella is to be emphasized.

Other important plants that illustrate the large grass type of stem are sugar cane and sorghum. Wheat, oats, and other small grains and field grasses illustrate the small hollow culm. The most easily available grass rhizome is that of quack grass, *Agropyron repens*. The hydrophytic monocots have interesting culms and rhizomes. Species of *Carex* having triangular stems, as well as round-stemmed species, and the cat tail, *Typha*, should not be overlooked. These subjects can be prepared by the methods outlined for maize.

Monocot stems of the nongramineous type may be obtained from several easily available plants. The trailing *Zebrina* grown in greenhouses has a soft stem that can be sectioned in paraffin. *Asparagus*

Fig. 13.8—*a*, Stem of hemp, *Cannabis sativa; b*, stem of basswood, *Tilia.*

sprengeri, an important plant in the florist trade, has a thin woody stem. The younger portions near the tip can be cut in paraffin, the old woody stems must be cut in celloidin. Wild species of *Smilax* have a woody stem. Kill the woody stems of *Asparagus* and *Smilax* in *FAA,* and cut in celloidin only if a sample embedded in paraffin cannot be cut.

DICOTYLEDONOUS AND CONIFEROUS STEMS

The apical meristematic regions of dicot stems have been discussed in the section on apical meristems. The tissue systems of these stems differentiate very rapidly close to the apex, and the first few internodes below the terminal bud show the fully developed primary tissues and the initiation of secondary activity. A convenient though artificial and arbitrary classification of stem types is in common use. *Herbaceous* stems develop comparatively little secondary wood, and, if a complete cylinder of wood is produced, it is laid down late in the growth period. *Woody* stems begin the formation of a complete cylinder of secondary wood early in the season and produce an extensive cylinder of highly lignified xylem. Every possible gradation of woodiness between these two types may be found in the plants about us. The following examples are recommended either because they are of economic importance or because they present some structural feature of fundamental importance.

Plants that can be grown quickly in pots are convenient subjects for the herbaceous stem. Plants of kidney beans, peas, and soybeans attain usable size in a short time. Actively growing field materials are the best source for sweet clover, alsike clover, and alfalfa (Fig. 13.7). Any of these legume stems can be killed in Craf III. The softer internodes can be carried through an acetone-xylene or alcohol-xylene series. The harder stems, especially soybean, cut better after dioxan or the butyl alcohols (Fig. 13.6).

The cultivated sunflower, *Helianthus,* and *Chrysanthemum* are good representatives of the Compositae. The common fleabane, *Erigeron,* is a suitable native subject in this family. The above stems seem to withstand the dehydrating action of *FAA* without marked plasmolysis, and a strong Nawaschin modification like Craf IV or V is satisfactory. The butyl alcohol process is recommended for these rather tough stems.

Bicollateral bundles are characteristic of the Cucurbitaceae and Solanaceae (Fig. 13.7). Important members of these families can be obtained easily. Seedlings of squash, pumpkin, or melon grow rapidly and furnish long hypocotyls as well as epicotyl materials. Do not use *FAA;* kill in Craf II, and use alcohol-xylene for tender stems and butyl alcohol for tough ones. Tomato and tobacco seedlings grow slowly, but they are almost indispensable subjects. Potato plants are easily grown from tubers. Stems of these plants are not killed properly by *FAA* but are preserved with excellent cellular detail in Craf II. Old, tough stems of tomato and tobacco must be processed in *TBA* or sectioned in celloidin. Small potato tubers, 3 to 6 mm. in diameter are easy to section. Kill in Craf I, and embed in paraffin by a slow, closely graded process. Longitudinal sections show that the tuber is a stem with an apical meristem which produces leaf primordia.

Medullary bundles and anomalous cambial activity occur in the Chenopodiaceae. The common weed *Chenopodium album* is probably the most readily available representative. Several related weeds are equally interesting. Kill in *FAA* or Craf III, and process in butyl alcohol or dioxan.

The foregoing methods recommended for specific herbaceous stems can be used with an extensive range of plants in many species of economic importance or academic interest. For instance, commercial fibers of primary and secondary derivation can be illustrated with the stem of *Cannabis sativa* (Fig. 13.8). As a broad general recommendation, use a mild chrome-acetic-formalin on tender materials, and process in alcohol-xylene or acetone-xylene. For moderately hard stems use Craf III, and for very hard stems use *FAA,* followed by dehydration and infiltration in dioxan or butyl alcohol.

The bush fruits like raspberry, blackberry, currant, gooseberry, and other plants having similar semiwoody stems may be handled like herbaceous stems while in the tender growing stages, but they eventually become too hard to process by the foregoing methods. Such hard materials usually must be handled like woody stems, as described in the following pages.

For the study of twigs of woody plants, material collected during the winter has some advantages. The previous season's xylem is fully lignified, secondary phloem is fully matured and firm, the cambium is clearly distinguishable as a layer immediately adjacent to the wood, and the cambium does not slip readily. However, if the development of cambial derivatives is to be studied, stems must be collected at intervals during the growing season. Such materials must be processed with greater care than dormant stems. Twigs should be taken to the laboratory promptly and cut into short pieces for killing as described in Chap. 2.

Many species of forest, orchard, and shade trees make excellent preparations for the study of young woody stems. The basswood, *Tilia* (Fig. 13.8 *b*), has become a great favorite, but there is no advantage in studying basswood in a region where it is not native. Species of *Populus, Fraxinus,* and *Acer* are easily sectioned. The apple and other fruit trees have been neglected as class materials, although they are easy to section. Tougher woods like oak, hickory, or locust are much more difficult to cut, and complete perfect sections are not obtained with such certainty. The standard coniferous subjects are the white pines, *Pinus strobus* in the east, and *P. lambertiana* or *P. flexilis* and several other five-needle pines in the west. These are representative of the five-needle or soft pines. For the hard pine type many more species are available, such as several species of yellow pine, the scrub pines, and jack pines. There is not much choice among the numerous two- and three-needle hard pines.

The principal American genera of conifers should be represented in a comprehensive stock of slides. Some of these trees are used as ornamentals, the commonest ones being *Abies, Larix, Tsuga, Thuja, Picea, Pseudotsuga,* and *Juniperus.* Shrubby conifers are among the commonest ornamentals, and specimens of shrubby species in the genera *Juniperus, Thuja,* and *Taxus* are readily available.

The methods of handling the woody dicots and coniferous stems are decidely stereotyped. The subdividing of such materials is illustrated in Fig. 2.2. The impermeability of the cork on woody twigs necessitates the use of a fluid of good penetrating powers, and *FAA* has long been the standard fluid. Stems may be left in this fluid for years. Preserved stems can be rinsed in several changes of 70% alcohol at 1-day intervals and sectioned without embedding. The celloidin method is recommended because of the ease and certainty of attaining high productivity by quantity production methods.

Woody stems having bark tissues are usually stained with the combinations recommended for herbaceous stems. Hemalum-safranin, safranin-fast green, and safranin-aniline blue have become standard stains. The method of handling sections and the staining processes are described in Chap. 8.

Transverse, radial, and tangential sections of the cambial region of woody plants make instructive preparations that are indispensable for a critical study of the three-dimensional aspects of cambium, the mechanism of abscission, and the structure of developing and mature elements of the xylem and phloem. The excessive use of transverse sections and the neglect of longitudinal sections build up an incomplete or even incorrect picture of the woody stem in the mind of the student.

Twigs are not satisfactory for making longitudinal sections in quantities. Unembedded twigs cannot be held in the microtome horizontally for longitudinal sections. If an embedded and blocked twig is sectioned longitudinally, only the outermost sections are strictly tangential, and only a few slices from the center are true radial sections, cut parallel to a ray. For first-class preparations cut accurately on the three desired planes, use blocks of wood and attached bark removed from living trees as illustrated in Fig. 2.2. Sectioning of such blocks is quite impossible without embedding in celloidin; whereas, with the celloidin method, perfect sections can be produced in quantities (Fig. 13.9 *b, c*). Collect the material in the winter when the cambium is firm. Soft wood like basswood, white pine, apple, or silver maple can be cut without special softening. Kill in *FAA,* and embed in celloidin. Hard woods like oak or locust must be treated with hydrofluoric acid after killing and hardening in *FAA.* The protoplasts cannot be expected to be in perfect condition after treatment in HF. The process is described in Chap. 8.

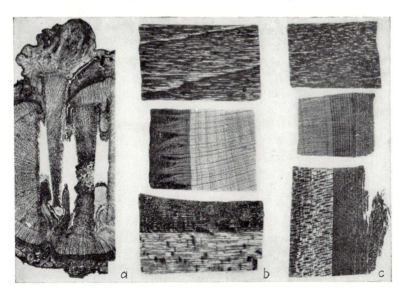

Fig. 13.9—Illustrations of material cut by the celloidin method: *a*, cross section of apple graft union; b, sapwood region of sector from 20-year-old trunk of *Tilia,* sections in three planes; *c*, sections from approximately 40-year-old trunk of apple tree. All subjects killed in *FAA*. The *Tilia* stem was infiltrated in Cellosolve solution of celloidin.

The desirability of using choice sections showing the bark in three planes cannot be overemphasized. In addition to serving as supplementary class material for studying the structure and development of the stem, such preparations serve as reference material for research, especially for pathological studies. Even in wood technology, in which the work is largely confined to the microscopic structure of the wood, preparations showing the cambium, phloem, cortex, and periderm are a valuable supplement.

Seasoned lumber is frequently used as a source of material for slides, and excellent preparations can be made from such material. However, parenchymatous elements such as xylem parenchyma and the epithelial cells of resin canals are collapsed and distorted. The preparations are adequate for diagnostic purposes and for the study of nonliving elements of the xylem. For best results, use properly seasoned wood and prepare the blocks so that sections can be cut accurately along the three conventional planes as described in Chap. 2. The specialized sectioning methods necessary for dry or hard woods are described on pages 84–85.

The Root

The processing of roots of seed plants for anatomical study is similar to the methods used for stems. The meristematic root tip is usually prepared by careful cytological methods; sections may then be stained either with a cytological stain for nuclear structures or stained with some histological combination. Batches of root tips that do not have abundant mitotic figures are usually set aside for histological preparations. The methods of obtaining root tips are described in Chap. 9.

The histogens of roots are evident at the root tip, especially if the preparation is stained to show cell walls as well as nuclei. The primary tissues are evident at the beginning of the root-hair zone, where the emerging root hairs can be detected with a hand lens. Initiation of lateral root primordia can be demonstrated at the upper limits of the root-hair zone, where the old root hairs are beginning to collapse. At this level the primary tissues are usually clearly differentiated, without being excessively woody.

Favorable subjects for illustrating the monocot root are maize and *Asparagus officinalis,* the garden asparagus. Germinate corn in

FIG. 13.10—*a,* Transverse section of root of *Asparagus officinalis* showing initiation of lateral root; *b,* brace root of *Zea.*

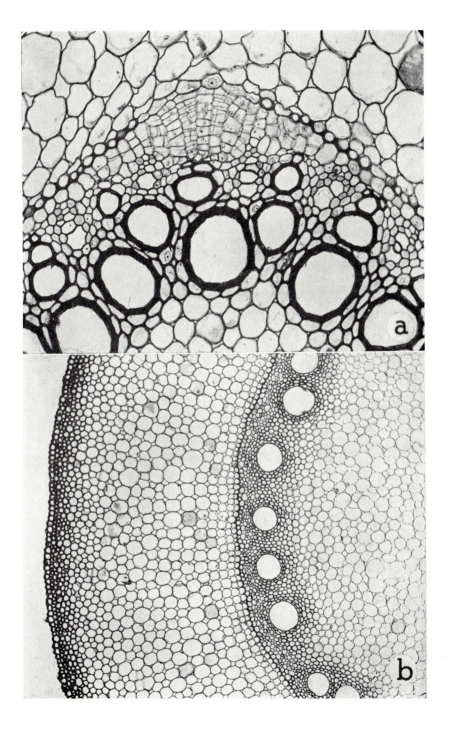

sphagnum (not in sand!), and remove pieces of root from the desired region. Roots grown in a moist chamber or water culture have excessively spongy, fragile cortical parenchyma, unlike the structure found in plants in a more normal environment. Tissues of the tip, the hair zone, and for some distance above, are fixed well in Craf II; the older tough roots must be killed in *FAA*. The brace roots from field-grown plants may also be used (Fig. 13.10). Roots of sugar cane, sorghum, and the small grains are processed by the above methods.

Asparagus roots are obtainable readily from volunteer seedlings that occur in the vicinity of asparagus beds. The softer portions of the root, within 3 cm. of the tip, are killed in Craf II, but older roots that have an impermeable hypodermis and endodermis must be killed in *FAA*. The butyl alcohol method is suggested for the harder pieces (Fig. 13.10).

Acorus calamus is almost a classical subject for the monocot root. This water plant is abundant in suitable locations, but material is often inaccessible, and the root has no advantages over *Asparagus*. The processing methods are identical for these two plants.

Smilax hispida root has a remarkably thickened endodermis, with laminated cell walls impregnated with brown coloring matter. Kill the roots in *FAA,* and try a batch with the butyl alcohol method, using celloidin if sections cannot be cut in paraffin. Safranin-fast green gives a brilliant contrast in which the prominent endodermis is reddish brown.

Young dicotyledonous roots are obtained readily from the large-seeded legumes, beans, peas, soybeans, and especially the horse bean, *Vicia faba*. The early stages, including the emergence of lateral roots, can be obtained from roots grown in a moist chamber of sphagnum. These roots do not become excessively spongy when grown in this manner. The older roots showing extensive secondary growth must be taken from plants grown in soil. Soybean and horse bean should be killed in Craf III. Young roots of the apple are particularly interesting because of the prominent Casparian strips in the endodermis. Material is not so easy to obtain as with plants that can be grown quickly from seed. Volunteer seedlings of apple can be dug up carefully and abundant roots of various ages obtained. Kill the youngest roots in Craf III and woody roots in *FAA*. Trial batches of older roots may

Fig. 13.11—Transverse sections of leaf of *Zea*: *a,* photographed at 14X with 48mm. Micro Tessar, reproduced at 28X; *b,* detail of trichome and blade.

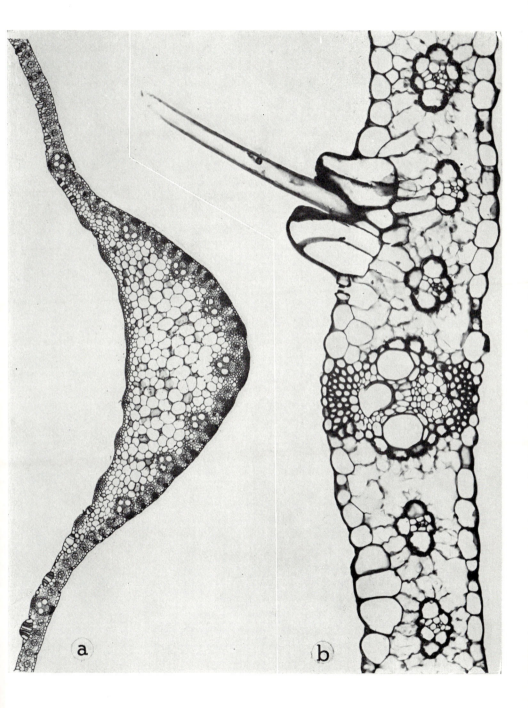

a

b

well be processed with tertiary butyl alcohol, and any age classes that cannot be cut in paraffin must be cut unembedded or in celloidin.

Flax root has a simple diarch stele. Roots can be obtained easily by germinating seeds in blotting paper. Radish, mustard, cabbage, and many other roots also may be grown in this manner to adequate size for primary tissues and processed by the methods given for apple.

Ranunculus root has long been popular as an example of the dicot root. The large fleshy roots of the buttercups, *R. septentrionalis* and *R. fascicularis,* are easy to obtain and to process, using the methods given for apple. The buttercup root is less likely to be of interest to the student than the roots of economic plants.

Large taproots like those of alfalfa, *Medicago,* and sweet clover, *Melilotus,* are handled like the older semiwoody roots of apple.

The tough, woody stems, rhizomes, and roots of ferns, horsetails and club mosses are most conveniently discussed at this point because they are handled like other woody materials. Collect the rhizomes of ferns in the spring, just after the fronds have fully expanded. Acceptable preservation of rhizomes can be obtained with *FAA,* and it is not improbable that for investigational work this formula could be adjusted to give good fixation with given species. For routine preparation of many species, uniformly good results have been obtained with Craf II. The subjects become very brittle after xylene, but very large rhizomes can be cut readily after *n*-butyl or tertiary butyl alcohols. It has been customary to embed hard rhizomes like those of *Pteris aquilina* in celloidin. However, the *TBA* process is satisfactory for portions of the rhizome that are not excessively hard, but have the woody structures adequately lignified to show the mature condition of tissues. The most brilliant and satisfactory stain is safranin with fast green. The presence of yellow deposits in the cells produces undesirable staining effects with the hematoxylins.

The fleshy root of *Botrychium* is recommended. Tests with *B. virginianum* have shown that Craf I gives much better fixation than does *FAA.* A very striking color contrast is obtained with safranin-fast green. Roots of the Boston fern and of available native ferns are processed as above.

Species of *Lycopodium* occur in abundance in some regions, and some species, especially tropical ones, are cultivated in conservatories. Stems may be fixed in *FAA* and carried through tertiary butyl alcohol to paraffin. It is usually necessary to soak the mounted specimen in warm water before sectioning. Roots are easy to process successfully by the same methods.

Selaginella has highly localized distribution, but excellent preserved material is obtainable from dealers, and several species are extensively cultivated in greenhouses. The processing is the same as for *Lycopodium,* the infiltration must be slow and thorough, because the stele is literally suspended in a highly parenchymatous cylinder and is easily torn in cutting.

The vegetative organs of *Isoetes* are studied only in advanced work in anatomy, and there is comparatively little demand for slides. Roots should be severed and divided into short pieces. The compact stem and rhizophore may be processed entire or quartered. The methods used for *Lycopodium* are satisfactory. The highly silicified stem of *Equisetum* has long been a problem for technicians. Penetration is difficult with an aqueous killing fluid, but *FAA* is satisfactory. The older stems must be desilicified by treating with hydrofluoric acid. Transfer directly from *FAA* to the acid diluted with twice its volume of 95% alcohol. After 2 days in acid, wash in 50% alcohol, making at least five changes at 4-hr. intervals. Observe the precautions concerning the use of HF given in Chap. 8. Dehydrate in *TBA,* and infiltrate slowly and thoroughly. Rhizomes and roots do not need to be desilicified, otherwise the processing is the same as for aerial stems.

The Leaf

The mesophytic dicotyledonous broad leaf is the type most commonly used for the study of the so-called typical leaf. Some general directions apply for the handling of most types of leaves. Leaves are easily damaged during processing by apparently minor mishaps. It is therefore desirable to kill duplicate batches in each of the formulas used, keeping one batch in the preserving fluid while the other one is embedded and tested. Consult Fig. 2.1 concerning the usual methods of subdividing leaves. Good results can be obtained with many leaves by killing in *FAA.* Soft leaves having small veins can be dehydrated in acetone or ethyl alcohol, whereas leathery leaves, or leaves with thick or wiry veins should be processed in butyl alcohol or dioxan. One batch of each subject may well be killed in *FAA* and another batch in one of the fluids given in the following specific recommendations.

The firm leaves of the trees and shrubs are represented by apple, cherry, rose, lilac, and privet. Leaves of apple and related plans may be killed in *FAA,* but occasional batches exhibit considerable plasmolysis (Fig. 11.1). Consistently good results can be obtained with Craf II and *TBA* dehydration. The latter reagent minimizes the brittleness of these subjects. A disadvantage of the rosaceous leaf is the presence of excessive brown pigmentation in the cell walls and masses of yellow gummy materials in the cells. The embedded pieces of leaf in the paraffin block are decidedly dark, and the staining effects tend to be muddy, especially with the hematoxylins. The use of a safranin-fast green or safranin-aniline blue combination makes slides with fairly clean color contrasts. Lilac and privet leaves can be processed by the above methods. Many other trees and shrubs have leaves in this firm-textured category. Geranium leaf is firm and easy to process. Craf V gives excellent results. Do not use pieces with large veins unless *TBA* is used for dehydration.

Leaves of softer character than the foregoing are illustrated by various easily obtainable legumes. Kidney bean, soybean, clovers, and alfalfa have more or less pubescent leaves; peas and horse bean have practically glabrous leaves. All of these leaves have been killed successfully in *FAA,* but failure occurs often enough to justify more critical methods. Excellent preservation of alfalfa and soybean leaf has been obtained with Craf III followed by an acetone-*TBA* series (Fig. 11.1). The thinness of the cell walls requires a stain of good contrast, such as hemalum, followed by safranin, the xylem stain. The coal-tar dye counterstains are likely to yield weakly stained parenchyma and barely visible plastids.

The leaves of the Solanaceae and Cucurbitaceae represent the very tender type of broad leaf. Subdividing of fresh leaves must be done with the greatest care because of the open and fragile construction of the parenchyma. The glandular hairs should also be preserved intact. The best killing is obtained with a mild fluid, such as Craf I. Practically perfect preservation of tobacco leaf has been obtained consistently with this fluid. Although the blade is soft in leaves of this type, the veins are large and firm, justifying the use of *TBA.*

Begonia leaf is an interesting tender leaf. The epidermal cells on both sides are enormous, the two layers occupying more than two-thirds of the thickness of the leaf. The narrow interior layer consists of poorly defined palisade and extremely loose spongy parenchyma. All these interior cells contain chlorophyll; each cell has relatively few, but very large chloroplasts. A leaf of this type is obviously difficult to preserve. Good killing has been obtained with chrome-acetic 0.5–0.5, washed by diffusion in a large volume of water, followed by the ethyl alcohol-xylene series. Cut 15 μ thick in order to keep the large epidermal cells intact. Coleus is another common greenhouse plant with soft leaves. They are preserved well by *FAA* and particularly well by Craf II.

The stereotyped construction of the mesophytic leaf permits the use of innumerable species to illustrate the type, making it possible to utilize plants that are readily accessible and characteristic of the region rather than use some classical species as if it had special virtues.

Most broad leaves are distinctly dorsiventral, the columnar palisade cells being on the upper or ventral side. Leaves that normally assume a vertical position do not have such distinctive palisade cells, the upper and lower tissue zones are nearly alike, and the dorsiventrality is obscured. The garden beet and sugar beet are good examples. These leaves can be fixed successfully in *FAA* or Craf III.

The study of the dicotyledonous leaf would be far from complete without a study of deviations from the typical mesophyte. Perhaps the most striking variations are the xerophytic adaptations. The leaves of species of *Dianthus* show a range from the relatively large, flat leaves of the greenhouse carnation to the waxy, narrow, rolled leaves of the rock garden species. These easily obtainable leaves are well preserved by Craf II. The tough cuticle becomes brittle after xylene but cuts well after *TBA* or dioxan. A brilliant stain is obtained with safranin-fast green.

Nereum oleander has leaves of unique xerophytic structure. The lower surface is indented with globose cavities or infoldings of the epidermis. Each pocket is lined with numerous hairs and contains many stomates. The upper epidermis is firm and highly cutinized. There are two to three layers of tough thick-walled hypodermal cells below the epidermis, and the deep-seated palisade cells are long and narrow. Killing fluids penetrate with difficulty. *FAA* is the most rapid of the satisfactory formulas but may cause slight plasmolysis. If the pieces cut transversely out of the fresh leaf are very narrow, not over 1 mm. wide along the linear dimension of the midrib, good penetration and fixation are obtained with Craf 0.30–1.0–5.0. Brittleness in paraffin is minimized by the use of butyl alcohol or dioxan. Safranin-fast green gives a brilliant and highly differential stain.

Leaves of citrus fruits are also of the leathery type and have an added interesting feature, the pear-shaped oil glands in the epidermis. The spongy parenchyma is compact and firm, and the palisade cells are small and closely spaced. The impervious character of the surface and compactness of the interior necessitate the use of *FAA,* which produces acceptable results. If the pieces of leaf are cut very narrow, Craf III produces excellent fixation.

Hedera helix is a remarkably efficient xerophyte that can with-stand severe drought. However, the leaf has no striking structural adaptations; its tissue organization is that of a stereotyped mesophyte. This very fact makes the leaf an interesting subject for comparative studies. Kill in Craf II.

The leaf of either of the common rubber plants, *Ficus elastica* or *F. pandurata*, is an interesting leathery, latex-bearing leaf. The small, compact epidermal cells are overlaid by a very thick cuticle. Under the upper epidermis there are two layers of large water-storage cells, under which there are two layers of small, short palisade cells. Two layers of compact hypodermal cells occur adjacent to the lower epi-dermis. The spongy parenchyma is very open and is transversed by the prominent latex vessels. The latex does not seem to be preserved in stainable form by *FAA*, but very well by the chromic acid fluids. Excellent results are obtainable with chrome-acetic 0.5–0.5 or Craf I.

Other illustrations of lactiferous leaves are easily obtainable. The leaf of the ubiquitous dandelion can be preserved in perfect condition by Craf I. Leaves of the common cultivated poinsettia and of the cultivated and native *Euphorbias* are well preserved by Craf I. These leaves are not brittle and may therefore be put through an alcohol-xylene or acetone-xylene series.

The *succulents* have very fleshy leaves that can be preserved well in Craf I and dehydrated carefully in normal butyl alcohol. The tissues are highly susceptible to damage, and a procedure that pro-duces severe distortion should not be condemned without a repetition of the process.

The gramineous leaf is represented by maize, sugar cane, sorghum, foxtail, and bluegrass. Corn illustrates well the border parenchyma of the vascular bundles (Fig. 13.11), but sorghum, and especially sugar cane, have more striking motor cells. Bluegrass is a good repre-sentative of the narrow type with prominent *bulliform* motor cells along the midrib. Foxtail is intermediate between the very broad and very narrow types. Acceptable preservation can be obtained with *FAA*, but for nearly perfect fixation use Craf III. This procedure has been repeated many times with corn, wtih uniformly good results. Prior to the introduction of the butyl alcohols and dioxan the older midribs of corn and other grass leaves were difficult to section without considerable breakage, but the use of these reagents has minimized the difficulty.

Monocotyledonous leaves other than the gramineous type should be included in a comprehensive collection. The leaves of lily represent the broad flat type. The extremely large stomates are the standard subject for studying sectional views of the stomate. Kill in Craf II and process by the alcohol-xylene method. The thinness of the cell walls necessitates a contrasting wall stain like hemalum. The stomates are shown with almost diagrammatic clarity. Leaves of *Zebrina, Rhoeo, Tradescantia, Polygonatum biflorum,* and *Smilacina racemosa* can be prepared by the above method.

The tender, cylindrical, hollow leaf of onion is preserved in excellent condition in Craf I and processed like those of lily.

The tough leaf of *Iris* is difficult to section. Fair cellular fixation can be obtained with *FAA* and excellent preservation with Craf I if the pieces are very narrow. The *TBA* process minimizes brittleness in paraffin.

The favorite subjects for the study of coniferous leaves are *Pinus strobus,* a soft pine, and *P. laricio austriaca, P. sylvestris,* or other hard pine. Needles collected in the late fall and in winter are very hard and become extremely brittle in the paraffin. The cells contain much granular resinous material which remains in the finished preparation. In July the needles are full grown, with all structural features fully developed, but they are still sufficiently soft to cut readily, and do not have excessive deposits in the cells. Kill in Craf III, and process in *TBA;* safranin-fast green yields a beautiful preparation. Longitudinal as well as cross sections should be made. Needles of the spruces (*Picea*) are also in the tough, wiry category and should be processed like those of pine.

The flat type of needle is represented by Douglas fir (*Pseudotsuga taxifolia*), the hemlock (*Tsuga*), or the fir (*Abies*). These may be killed in the fluids recommended above. Although these relatively soft needles can be processed through xylene, dioxan and butyl alcohol improve the cutting properties.

The broad leaf of *Ginkgo biloba* should not be omitted from a study of the gymnosperm leaf. Collect leaves in July, and kill in Craf III.

The leaves of cycads are difficult subjects because of their tough, xerophytic features. Select pinnules that are not fully matured and toughened. Cut transversely into narrow pieces and kill in *FAA.* The nonalcoholic fluids penetrate poorly, but excellent fixation in small pieces is obtained with Craf II. Xylene renders the tissues very brittle, but tertiary butyl alcohol permits satisfactory sectioning.

Fern leaves are readily obtainable from the common Boston fern. Use pinnules that have expanded to maximum size but are still bright, shiny green. Old leaves contain discoloring deposits in the cells. Kill in *FAA* or Craf II. Other conservatory or native ferns may be prepared by the same methods.

The foregoing recommendations dealt with mature leaves. Advanced students are invariably interested in the development of the leaf. The place and mode of origin of leaf primordia and the early stages of leaf development are evident at the growing points of stems, and the processing of suitable materials is discussed in the section dealing with the stem.

14. Thallophyta and Bryophyta

This chapter brings together plants and plant organs that require essentially similar treatment. For instance, a segment of mushroom cap, a liverwort thallus and a moss gametophyte may well be prepared for sectioning by identical methods. Filamentous algae, free-floating filamentous fungi and moss prothalli also present similar problems. Fungi that live within tissues of higher plants will be discussed in terms of the responses of the fungus and the host to processing.

Algae

The most satisfactory method of studying algae is by the use of fresh living material in conjunction with well-preserved bulk material. Except for some critical cytological features, most of the life history can be worked out without stained preparations. Stages that have short duration must be preserved when available and subsequently studied from temporary mounts. However, permanent stained slides are indispensable for research and have a legitimate place in teaching to supplement bulk material. The methods of processing the most commonly used algae are outlined briefly in this chapter, with frequent references to the whole-mount methods in Chap. 10.

GREEN ALGAE

These plants exhibit a wide range of size, complexity of organization, and habitat. The following simple precautions should be observed in collecting, transporting, and storing plants:

1. Keep the plants in their natural substratum (water, soil, bark of tree) until the moment of killing.

2. Avoid subjecting the plants to excessive heat or to desiccation during storage or transportation.

3. Unless culture methods have been carefully worked out, kill the plants as soon as possible after collecting.

4. Subdivide or spread out large masses of material to promote rapid killing and hardening (fixing).

5. Keep intact the organization of the filament, or other type of colony.

The preservation of green algae for bulk material and for permanent stained slides is treated at some length in the chapter on whole-
mount methods. The advanced worker will find further details in
Johansen's (1940) comprehensive treatment of culture methods and
processing of this group.

BLUE-GREEN ALGAE

These algae have such simple cellular and colony organization and
are so easy to study in temporary whole mounts that the use of prepared slides is less justifiable than with any other group of thallo-
phytes. Fresh cultures are easily found in a wide range of habitats;
in stagnant pools, tanks, barrels, and crocks, on potted plants with
stale soil, on damp, poorly drained soil, and innumerable other places.
Some forms like *Oscillatoria, Rivularia, Nostoc,* and *Gloeocapsa* may
be found in masses that are practically pure cultures. Collections of
such materials are easy to preserve. Because of the dense undifferen-
tiated character of the protoplast the crudest methods of preservation,
such as 5% formalin, may be used. If the reagents are available, one
of the fluids containing glycerin should be used. Temporary or permanent whole mounts can be made as described in Chap. 10.

THE MARINE BROWN AND RED ALGAE

The algae in these groups are available in fresh condition for a
very limited number of schools. Large quantities of these plants are
used by schools that are totally dependent on outside sources for their
materials. Therefore, a detailed discussion of methods of collecting
and preserving these plants would have but limited usefulness. For
occasional casual collecting on one's travels, the simplest preservative
is 5 to 10% formalin in sea water. Further refinements are the addition of 5 to 10% glycerin and ½ teaspoonful of borax to 1 liter of
fluid.

For more critical preservation, nothing has been found to excel
chrome-acetic, with or without addition of osmic acid. One of the
best formulas is the Chamberlain formula (Table 3.1) made up with
sea water. Subsequent processing of filamentous forms for whole
mounts is outlined in Chap. 10.

If materials are purchased from collectors, the purchaser should
indicate whether the material is to be used for temporary slides or
to be processed for permanent preparations. Several reliable collectors
will furnish material in specified stages of the life history, carefully
fixed in a suitable fluid determined by the collector or specified by

the purchaser. Such materials will yield excellent preparations by the methods recommended in Chap. 10.

Few of the algae are microtomed for making slides. Some selected items that are customarily microtomed are discussed briefly.

CHARA AND NITELLA

The growing points of *Chara* and the sex organs of mature plants must be sectioned to show cellular organization and nuclear structures. Kill in medium I chrome-acetic or Craf II. These fluids contain enough acid to remove much of the troublesome incrustation. The condition of the material after 1 week in the fluid can be easily ascertained by examining a whole mount. If abundant material is available, several variations of these formulas should be tried, and the batch having the best fixation used for embedding. Older oögonia and zygotes are not readily penetrated by the above fluids; *FAA* should be used.

These plants become very brittle in xylene, but they section satisfactorily after the butyl alcohol process. Examine small samples during the process, thereby saving further work if a batch has undergone plasmolysis. Infiltration should be gradual, with the time interval in the oven reduced to 2 days or less. The staining of different batches is highly variable. Try iron hematoxylin and safranin-fast green.

Fucus and similar bulky forms are usually sectioned to show gametangia. Kill in medium II or strong chrome-acetic made up with sea water. Dissect out some of the gametangia to ascertain which fluid preserves them best at the given stage. Process in *TBA* or dioxan. Sections are difficult to affix to the slide. It may be necessary to use an alcoholic bulk stain with some batches. Brilliant staining of immature sperms in the antheridia of *Fucus* has been obtained with iron hematoxylin; sharp staining of nuclei during cleavage in the oögonium is very difficult.

The more massive Rhodophyceae that cannot be satisfactorily made into whole mounts may be sectioned by the method given for *Fucus*. Since the great majority of readers do not have access to fresh plants, the purchase of carefully preserved material from biological supply firms is recommended.

Fungi

The processing of fungi involves many problems that are common to other categories of previously described subject matter. For example, in processing a fungus parasitic on a leaf, the tissues of the host must be preserved unchanged; the fungus, with an entirely different chemical and physical make-up, perhaps an alga-like siphonaceous plant body, also must be preserved intact. Another task may involve cutting a tough piece of wood bearing a delicate plasmodium, presenting a conflict between the need for drastic methods and refined methods. In order to minimize duplication of procedures in this chapter, it is proposed to use extensive cross references to appropriate sections of the text and to give detailed directions for procedures that are not adequately covered elsewhere in the manual.

SCHIZOMYCETES

The preparation of slides from cultures of bacteria is described in detail in textbooks of bacteriology. The bacteria are discussed in this manual only in conjunction with a host plant. A few typical examples of plant tissues and their bacterial invaders will illustrate the general methods of processing. *Bacterium stewartii* invades the vascular system of corn, forming a shiny yellow mass in the xylem elements. Because of the virulence and ease of dissemination of the disease, it is unwise, in regions where the disease is not normally

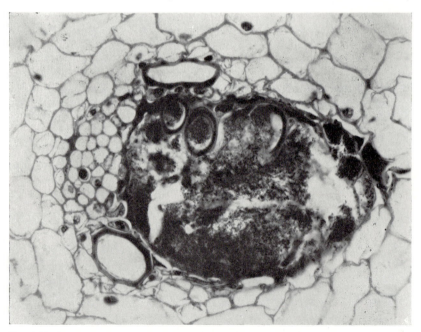

FIG. 14.1—Vascular bundle of *Zea*, with bacterial mass (*Bacterium stewartii*) in the protoxylem.

present, to infect plants to obtain diseased tissues. Preserved tissues can be purchased from the supply houses. The most satisfactory killing fluids are *FAA* and *FAA*-bichloride of mercury (Chap. 3). Chromic acid seems to become fixed in the gelatinous bacterial slime and interferes with clear staining. Process the corn stem or leaf as described in the section dealing with vegetative organs of seed plants (Chap. 13). Iron hematoxylin gives a brilliant differentiation of the bacteria. The xylem may be lightly stained with safranin or gentian violet, but the slime between the bacteria must be thoroughly destained (Fig. 14.1).

The above methods are satisfactory for the preparations of cucumber stems infected with the wilt organism, *Erwinia tracheiphila,* and succulent leaves and twigs of apple or pear infected with *E. amylovora,* the fire blight organism.

MYXOMYCETES

The slime molds are customarily studied from living cultures of the slimy plasmodium and from dried specimens of the fructification. These spore cases are exceedingly delicate and beautiful objects. Sporangia that are nearly mature can be mounted into permanent slides. Transfer directly into 95% alcohol for 10 min. Pass through three grades of anhydrous alcohol-dioxan at 10-min. intervals, then into pure dioxan, and mount in thin dioxan-balsam. A similar butyl alcohol series may also be used. The nuclei are exceedingly small, and microtoming and staining are tasks for the experienced cytologist.

PHYCOMYCETES

The saprophytic members of this group should be studied in culture whenever possible, and the use of prepared slides should be discouraged. Stages that are of short duration or difficult to obtain can be preserved, for either bulk material or permanent slides. The representatives of this group presented below are in common use for teaching, and the process for each plant has been thoroughly tested and may be regarded as type processes applicable for similar subjects. Strict taxonomic sequence is not maintained in the following discussion; organisms that are processed by similar techniques may be discussed simultaneously.

Zygomycetes.—The order Mucorales contains the best-known members of this group. Species of *Rhizopus* and *Mucor* are easily grown in culture and studied to best advantage from whole mounts. Developing and mature zygospores can be preserved by cutting out selected pieces of the culture agar and killing in *FAA,* which also serves as a storage fluid. Such material can be used for mounts in water or lactophenol or for excellent stained permanent preparations of whole mounts (Chap. 10). Cytological preparations require such highly specialized and almost specific methods that the ambitious student should study the research publications of a given species for details of procedure.

Oömycetes.—Plasmodiophora brassicae is parasitic in the roots of cabbage and related plants. The plasmodium can be demonstrated in young roots that are just beginning to undergo distortion. Stages of cleavage and spore formation are obtained from increasingly gnarled and distorted roots. *FAA* gives good fixation, but Craf III is superior. A simple hemalum-safranin stain is adequate for most purposes, safranin-fast green is more contrasty, and iron hematoxylin gives the most brilliant differentiation of the parasite nuclei.

Synchytrium decipiens, parasitic on the hog peanut, and related parasites yield striking preparations, but poor fixation is frequent, and sectioning is unproductive, making the slides somewhat expensive. Craf II was found to give excellent fixation. Iron hematoxylin is by far the most satisfactory stain.

Saprolegnia and allied water molds are readily obtained and easily cultured, furnishing abundant vegetative and sexual material for study in the living condition. The best sources are dead fish and water insects, or steam sterilized house flies placed into a large crock of pond water. Whole mounts can be prepared by the general methods given for filamentous plants (Chap. 10). Ascertain the correct killing formula for the species being studied by testing small masses in a weak chrome-acetic or Craf I and manipulating the acetic (or propionic) acid content. Whole-mount and sectioning methods are used for cytological study, and the reader is referred to research publications for these highly specialized and difficult methods.

Pythium and *Phytophthora* are most effectively studied in culture, but preparations can be made by the methods suggested for the water molds.

Albugo (Cystopus), the "white rust," is an indispensable subject in teaching. Several species occur on common crops and weeds. For the conidial (zoosporangial) stage, select pustules that have just ruptured. Fully opened pustules will have most of the spores washed out during processing. The sexual stages arise after the conidial stage is on the decline, and the host tissues show evidence of hypertrophy. *A. candida* and *A. bliti* are fixed in perfect condition in Craf II.

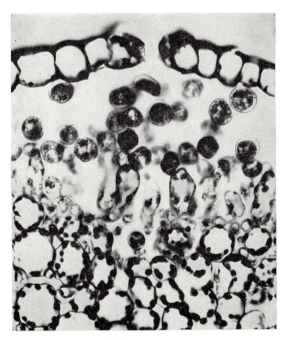

Fɪɢ. 14.2—Portion of pustule of *Cystopus candidus* on stem of *Capsella,* Craf 0.30–
1.0–10.0, acetone-tertiary butyl alcohol.

Safranin-fast green and hemalum-safranin are both excellent
combinations for elementary use, and iron hematoxylin is a good
nuclear stain (Fig. 14.2). Mature oöspores are abundant in the older,
hypertrophied stems and fruits of the host, therefore, a more vigorous
killing fluid is needed; *FAA* and standard Nawaschin are both
satisfactory. As might be expected, the host cells are in a distorted
condition, a fact not recognized by inexperienced teachers.

Peronospora parasitica, a common parasite on crucifers, yields
excellent preparations by the methods described for *Albugo.* The best
compromise fluid for preserving both the host cells and the fungus is
Craf II. Safranin-fast green differentiates the nuclei of the fungus,
but not so sharply as iron hematoxylin.

ASCOMYCETES

The mold members of this group, such as *Aspergillus* and
Penicillium, are so easy to culture and study from wet mounts that
permanent slides are seldom necessary. Permanent slides can be made
by the common whole-mount methods. Such slides are useful for

quick reference rather than for detailed study. The production of the ascigerous stage is highly uncertain, and best studied from whole mounts.

The Erysiphales are of interest because many members are parasites of considerable economic importance. *Erysiphe graminis* occurs on many grasses, from which the conidial stage is easily obtained in abundance. The conidia are studied best by freshening detached leaves in a moist chamber and examining the surface under moderate magnification, with oblique surface illumination. Microtome sections are indispensable, however. Longitudinal sections of the leaf show the clusters of long, finger-like haustoria in the long epidermal cells. Select leaves that are young and soft, kill in Craf III, and process like any young grass leaf. Iron hematoxylin is the most desirable stain, although a *good* triple stain is indeed beautiful.

Other interesting or important species are *Erysiphe polygoni* on the common weed *Polygonum aviculare, E. humuli* on the rose, *Podosphaera oxycanthae* on cherry, *Microsphaera alni* on lilac, and *Uncinula salicis* on willow. For the best slides of haustoria, collect in the conidial stage. Development of the perithecia can be studied from successive collection up to the stage in which the perithecia begin to turn gray. Kill in Craf III if the host cells are to be preserved, or in Craf I for good preservation of the young perithecium. The latter fluid does not seem to serve so well for the host cells. Stain as recommended for *Erysiphe graminis*. Mature perithecia are very brittle and difficult to section. Furthermore, this stage is studied to best advantage from dissections and macerations of bulk material preserved in one of the fluids in Chap. 10. The sectioning of the decayed, brittle, overwintered host leaves is a thankless and pointless task except for research.

Members of the Pezizales are of considerable cytological interest as well as economic importance. *Sclerotinia fructigena* occurs on cherries and plums. The conidial stage need not be sectioned, and sections of the hard sclerotia are not particularly interesting. The delicate goblet-like apothecia yield excellent sections. Very early in the spring look for the apothecia arising from mummified fruits. Collect cups of various sizes and preserve each in a separate vial of Bouin's fluid or Craf I. Successive stages of ascospore formation will be obtained in this way. Longitudinal sections through the center of the cup show numerous perfectly aligned asci and ascospores. Use iron hematoxylin for nuclear details, but safranin-fast green shows both nuclei and trama very well.

Pyronema confluens is common on steam-sterilized soil in the greenhouse. Cytological preparations of the sex organs are a task for the skilled investigator, and the reader is referred to the research literature. The apothecia are processed like those of *Sclerotinia*.

Pseudopeziza medicaginis is parasitic in the leaves of alfalfa and other legumes. Collect material when the pustules are just opening and the apothecia are bursting through the epidermis. Excellent preservation is obtainable with *FAA*, and staining presents no difficulties. Even the simple hemalum stain differentiates the ascospores.

Sarcoscypha coccinea has a brilliant red, dainty, cup-like ascocarp that can be killed entire and processed exactly like *Sclerotinia*. The larger cups like those of *Peziza repanda*, *Urnula*, and the familiar ascocarp of *Morchella*, the sponge mushroom, should be suitably subdivided and processed as above.

Fleshy portions of the fructifications of other Ascomycetes are handled like the foregoing types. Interesting slides are obtainable from *Hypomyces*, a parasite on mushrooms; *Cordyceps*, parasitic on insects; the fruiting head of *Claviceps*, the ergot fungus; the saprophytic *Neurospora*. Species of *Nectria* in which the stroma is moderately soft can be sectioned. Remove the stroma down to the wood, subdivide vertically into narrow strips, kill in *FAA* or Craf II, and process in *TBA*. Always examine freehand sections or smears of fleshy Ascomycetes to determine whether the desired stage of ascus formation is present.

In the Taphrinales (Exoascales) only the genus *Taphrina (Exoascus)* is of importance. *Taphrina deformans*, the casual organism of peach leaf curl, is very abundant in some localities. The malformed succulent leaves are well preserved by Bouin's solution or Craf II. Sectioning and staining present no difficulties.

Venturia inaequalis, the apple scab organism, is widely distributed and easily obtainable in the conidial stage on the leaf. A vigorous fluid like *FAA* or Craf V is necessary. A simple stain such as hemalum-safranin is adequate. The perithecia mature in early spring on last year's decayed leaves. Such material can be studied well from newly gathered soaked leaves or bulk-preserved leaves. Such material yields permanent slides of decidedly ragged appearance, and microtoming is therefore to be discouraged.

BASIDIOMYCETES

This group contains a great diversity of forms and involves a wide range of techniques. We are again confronted with saprophytes that can be detached from the substratum and processed easily, whereas the parasitic members require adequate preservation of both parasite and host. In the following discussion the taxonomic order is subordinated to methods of preparing the material.

Ustilaginales.—These parasites occur on a wide range of hosts, but the most interesting members occur on important crop plants. *Ustilago zeae,* the corn smut, is found on all aerial parts of the corn plant. Smut galls on stems, leaves, and ovaries should be collected. Very young smut galls, that have not markedly distorted the organ being attacked, show the host cells in good condition, and contain active and rather sparse mycelium. Kill this stage in Craf III. Older galls having a milky white interior contain a great mass of mycelium and distorted host cells. Small pockets of chlamydospores occur in the white mass. This stage, which can be ascertained by freehand sections, is the latest useful stage. Kill these older galls in *FAA*.

The mycelium of corn smut has a strong affinity for hemalum, and a simple hemalum-safranin stain shows the hyphae stained blue-black, chlamydospores stained red, the thin walls of the host cells stained blue, and the lignified elements stained red. Use iron hematoxylin for nuclear studies.

Other common smuts, such as *Ustilago levis, U. hordei, U. avenae,* and the bunts, like *Tilletia tritici,* can be processed by the above methods.

The chlamydospores of many smuts and bunts germinate readily in water or carrot decoction. The promycelia and sporidia are studied best from wet mounts from culture, but the material can be made into permanent mounts by the dioxan or butyl alcohol methods (Chap. 10).

Uredinales.—The rusts rank among the most destructive and widespread plant pests, and class materials illustrating the important phases of the life cycle of the rusts are indispensable. The story of wheat rust has been so well publicized that the organism may well be the standard item representing this group.

Puccinia graminis has its red uredinial and black telial stages on wheat and many other grasses. The red summer-spore stage occurs on young leaves and is therefore easy to section. The black winter-spore pustules occur on older leaves and on the stems, both of which are difficult to section without tearing. Use the youngest leaf showing the telial stage, avoiding the use of stem material if possible. Kill in *FAA* and process like any leaf parasite. The pycnial (spermogonial) and aecial stages on barberry occur on young, tender leaves that are preserved fairly well by *FAA,* but Craf I followed by careful embedding yields superior results. For the most critical cytological requirements, use the Flemming modifications as described in research papers. Many stain combinations give excellent results for class material, safranin-fast green is particularly good, but iron hematoxylin is by far the best as a nuclear stain.

P. coronata, the crown rust, is probably second to wheat rust in importance. The uredinia and telia on *Avena* and other grasses and the pycnia and aecia on *Rhamnus* (buckthorn) are treated like wheat rust.

Two common species of *Gymnosporangium* have the telial stage on *Juniperus,* producing woody galls of stem tissue in which the mycelium is perennial. The younger galls are soft enough to be sectioned in paraffin. Divide into wedge-shaped pieces, kill in *FAA,* and process in butyl alcohol. The pycnia and aecia of *Gymnosporangium juniperi-virginianae* occur on *Pyrus,* and those of G. *globosum* on *Crataegus.* Treat like the aecial stage of wheat rust.

Melampsora is very common on willows and poplars. The bright yellow uredinia, which may entirely cover the leaf, are handled like other leaf rusts. The coal-tar dyes do not seem to be so selective for nuclei as iron hematoxylin. The telial stage on the old leaves is a difficult problem, the host cells become very brittle in paraffin, and the nuclear staining is selective only with iron hematoxylin.

A great diversity of host tissues in which rusts are found necessitates more or less specific adjustment of the killing fluid for each problem. The foregoing recommendations are based on successful preparations and will serve as a guide for other problems in this group.

Tremellales.—The order is characterized by the small, gelatinous fructifications. Septation of the basidium differs in the several families, and some authors regard as orders some of the families incorporated here. The delicate fruit bodies must be collected in an absolutely fresh condition or the time spent in processing them is wasted. The portion near the substratum is of no interest; remove the substratum completely, and kill the entire or subdivided fruit body in weak chrome-acetic, or in Craf I. Exercise extreme care during dehydration and embedding. The best stain is iron hematoxylin, with safranin-fast green as second choice.

Agaricales.—The primary consideration in the processing of this group is to maintain intact the more or less exposed, delicate basidia and especially the exceedingly fragile sterigmata on which the basidiospores are borne. The texture of the trama of the fruit body ranges from the soft, fragile pileus of a small *Coprinus* to the "woody" perennial pileus of *Fomes*. The softer members are difficult to preserve in normal condition but are easy to section, whereas the leathery fructifications can withstand processing but are very difficult to section.

The basidia of many species of Agaricaceae have been successfully fixed in weak chrome-acetic, in Craf III, or in Allen-Bouin II and III. The last is particularly good for cytological details. Bouin's solution has given good results, but it is rather erratic. Dehydrate in alcohol or acetone, beginning with 5% and using steps of 5% at 15- to 30-min. intervals. Iron hematoxylin and gentian violet-iodine are excellent for nuclear details. Safranin-fast green stains the nuclei well enough and also shows the gill and trama structure.

Softer members of the *Clavariaceae, Hydnaceae,* and *Polyporaceae* are processed as above; the leathery and woody forms must be dehydrated in butyl alcohol or dioxan. Fortunately, basidia mature in the soft new growth in even the toughest perennials.

Exobasidium occurs on *Vaccinium, Rhododendron,* and other members of the heath family. Kill in *FAA* or Craf III. Because of the leathery texture of the host the use of butyl alcohol is advisable.

FUNGI IMPERFECTI

This category includes fungi for which the perfect or sexual stage has not yet been found. The perfect stage, when discovered, is found to be a basidial or ascigerous stage, and the organism is then transferred to the appropriate group. Sporulation is by conidia, produced either at random on the mycelium or in closed pycnidia. The vegetative mycelium may be a superficial saprophyte, a saprophyte within dead tissues, or a parasite within tissues.

Mycelium and conidia from cultures can be prepared as whole mounts by the general methods outlined in Chap. 10. Parasitic species are handled in accordance with the properties of the organ on which they occur. Leaf parasites are the easiest to handle. The following illustrations are selected from successful preparations of important fungi.

Diplodia zeae grows readily in agar culture and produces abundant pycnidia. Cut out small pieces of agar bearing the pycnidia, fix in Craf I, and embed in paraffin. A heavy overstain in hemalum, slightly differentiated in HCl, stains the hyaline portions of the fungus very well. The pycnidia and the mature spores have considerable pigmentation.

Cercospora beticola is common on garden and sugar beets. Excise the youngest lesions to obtain sections embracing healthy tissues as well as diseased areas. If material must be killed in the field where an aspirator is not available, use *FAA,* which gives adequate fixation. Excellent preservation can be obtained with Craf III. Iron hematoxylin and safranin-fast green are the preferred stains.

The wood- and bark-inhabiting pycnidia are handled like perithecia of similar habitats. Such resistant subjects must be killed in *FAA,* and butyl alcohol is the preferred dehydrant.

LICHENS

The lichens are found in a wide range of habitats, from the mist-soaked rocks under a waterfall to the sun-baked face of a boulder. Collections should include a portion of the substratum whenever possible. Specimens usually are dried, and stored in containers that prevent breaking of the fragile dry plant. If wet preservation is preferred, use one of the fluids given in Chap. 10. Microtome sections of the vegetative thallus have little justification. The association of the green algal cells and the fungal mycelium is shown best by dissections and freehand sections of fresh or preserved material. The ascocarps should be preserved in fluid, examined with a hand lens for general organization, and teased apart for examination of bits of the hymenium under a microscope.

Microtoming of the ascocarp is a vexing problem with most species. The gelatin in the plant body becomes dry and brittle, and the sections fail to ribbon and do not adhere well to the slide. Select a species with a small, shallow cup-like apothecium. Kill in *FAA* and dehydrate in butyl alcohol. Soak the embedded blocks in warm water before sectioning. Staining presents no difficulties if selectivity for the diverse components is not demanded. Safranin-fast green is probably the best simple combination.

Bryophyta

The liverworts and mosses have such wide distribution and range of habitat that some representative member of the group is usually available for study. The most common liverworts are the aquatic *Riccia,* the well-known *Marchantia,* and two rock-inhabiting species, *Conocephalum conicum* and *Reboulia hemisphaerica. Anthoceros* seems to be less common, but it is easily overlooked if sporophytes are not present. Large and conspicuous mosses are usually preferred, the best-known ones are in the genera *Polytrichum, Mnium, Catherinia, Funaria, Rhodobryum,* and *Sphagnum.* Liverwort and moss species that are not locally available can be purchased from supply houses, preserved either for bulk specimens or for sectioning, as specified by the purchaser.

These fragile plants must be collected and handled with care, taking precautions to keep the plants moist and undamaged until the time of killing. Entire plants preserved in fluid are indispensable for teaching. The most useful bulk preservatives are described in Chap. 10. Preservation and processing for embedding must be carried out with painstaking care, approaching cytological methods.

Hepaticae

The following recommendations, based on *Marchantia,* will apply to a wide range of liverworts. The young, actively growing thallus is usually sectioned to show the construction of the pores, the highly spongy chlorenchyma, and the gemmae. Cut out 4-mm. squares of tissue. Gemma cups should be excised with a small square of thallus. Antheridial and archegonial receptacles should be collected when they are just beginning to be elevated above the thallus. The gametangia are at their best at this stage. When the archegonial disk has been fully elevated, make a collection for the developing sporophytes. A complete series of developmental stages may be obtained by collecting at intervals. Kill in weak chrome-acetic or Craf I. A closely graded alcohol-xylene series is recommended.

The thickness of sections can be judged best at the time of microtoming. Examine a few trial sections by melting the ribbon on a slide, and decide whether the trial thickness includes the desired structures and is sufficiently thin to show internal detail. Sections will range from 6 μ for a careful examination of young antheridia, to 15 μ for maturing capsules. A simple hemalum stain, with perhaps a light counterstain of erythrosin, sets off all essential structures very well. A multicolor stain combination is quite pointless. Iron hematoxylin is the ultimate choice for cytological details.

Gemmae can be studied conveniently by dissecting them from gemma cups of fresh or preserved thalli. Permanent whole-mount slides of gemmae are of little value, but such mounts can be made by the methods outlined for preparing filamentous green algae. Microtome sections are necessary to show the initiation and development of gemmae.

Riccia and *Anthoceros* are somewhat more difficult than the foregoing type, because the sex organs are sunken in the thallus. Skillful freehand sectioning reveals the presence of sex organs and eliminates the fruitless sectioning in paraffin of many vegetative thalli. The developing sporophytes of *Riccia* are visible within the thallus, and various stages can be classified roughly by size. Remove enough of the thallus with these organs to show some of the enveloping cells. Fruiting thalli of *Anthoceros* should be killed entire, in a vacuum jar (Fig. 3.1), and the sporophytes dissected away with a section of thallus after hardening in the fluid for several days. Both transverse and longitudinal sections of the sporophyte should be made.

The leafy liverworts are easily overlooked on collecting trips, and therefore do not receive adequate attention. *Pellia* and *Porella* are most commonly used to illustrate this group. They can be processed like the mosses as outlined below.

Musci

These plants are readily obtainable in fresh condition during the greater part of the year in all but the most severe climates, and they can be grown easily. They make usable dried specimens and can be preserved in excellent condition in the fluids given in Chap. 10. Gemmae, fully developed sex organs, and most features of the capsule can be studied from dissections. Prepared slides are needed principally for studying young sex organs, gametes, and some features of the developing sporophyte.

For the study of sex organs the large and more common species of *Mnium, Polytrichum,* and *Rhodobryum* are recommended. The proper killing fluid for sex organs and gemmae of mosses and leafy liverworts can be determined quickly. Obtain fresh turgid plants, dissect out a few short pieces of the shoot bearing the sex organs, and immerse in the fluid that is to be tried. Exhaust the air that adheres tenaciously among the leaves. After 1 hr. in the killing fluid dissect out a few gametangia, mount in a drop of the fluid, and examine with a microscope under at least 400×. If plasmolysis has occurred, adjust the formula. It is a good practice to try *FAA* and *FPA* (page 15). If these cause excessive shrinkage, try Craf I, an excellent formula. Adjustments in this formula are made by increasing the ratio of acid until no marked plasmolysis occurs. Use the stains recommended for liverworts.

Capsules of mosses are a vexingly difficult subject. Young green capsules of *Mnium cuspidatum* and *Funaria hygrometrica* are penetrated by Craf I, but for older, coloring capsules, *FAA* or *FPA* must be used. However, the interesting stages of sporogenesis take place long before the capsules become brittle, and there is little need for slides of old capsules. The dehydrating must be gradual, and *TBA* is preferred. The embedded capsules should be oriented carefully in the microtome, and both longitudinal and transverse sections are desirable. The capsule has enough internal differentiation to justify the use of a triple stain; however, the simple combinations given on Staining Charts II and III are usually adequate.

The sporulating capsules are studied to best advantage either from fresh plants, wet-preserved plants, or dried specimens from which they can be removed and thoroughly soaked in water or lactophenol (Chap. 9). Spores can be germinated readily and the protonema held at any stage by refrigeration under weak illumination. With a little planning by the instructor there is little excuse for using permanent prepared slides of protonema, although these can be made by the methods used for delicate algae.

15. Reproductive Structures of Vascular Plants

The preparation of vegetative organs of the highly diverse members of the phylum Tracheophyta, the vascular plants, is discussed in a separate chapter because such organs require similar techniques (Chap. 13). Similarly, the reproductive organs of the Tracheophyta present common problems of processing and staining and are therefore brought together in the present chapter. *Orders* as well as *Classes* ar used as major headings.

Lycopodiales

The organization of the strobilus of the club mosses should certainly be studied by dissection, and there is no point in embedding entire strobili. Ascertain the stage of sporogenesis in each cone by dissecting out a sporangium and crushing out the contents. Mature sporangia containing dry, hard, brittle spores should not be embedded unless a cytological study is to be made. Subdivide the cone transversely, and kill in *FAA,* medium chrome-acetic, or Craf III. Sections should be stained in safranin-fast green or iron hematoxylin.

The gametophytes of these plants are exceedingly rare, although they are said to occur in abundance in localized areas. Gametophytes may be purchased preserved in *FAA.* The soft thallus is easily sectioned in paraffin and stained.

Selaginellales

There is very little justification for making sections of the strobili of these plants because dissections under a binocular reveal so much more of the orderly organization of the cone. Dissected and crushed sporangia likewise present a three-dimensional picture that is lacking in sections. The study of nuclear details of sporogenesis and the development of gametophytes within the spores is a task for the experienced investigator.

Isoetales

Young sporangia of *Isoetes* arise on small sporophylls closely appressed to the rhizophore. Dissect away the sporophylls under a binocular, trim off much of the sporophyll, and kill the sporangia in medium chrome-acetic or *FAA*.

The large sporophylls and mature sporangia are inadequately represented by sections. To section mature spores, cut the sporangia away from the sporophylls, drop the ruptured sporangia into a centrifuging tube of *FAA*. Process in butyl alcohol, centrifuging the mass after each change. Much tearing of the spore wall can be expected during sectioning.

The preparation of gametophytes should be undertaken only after a study of research literature in which methods are given for germinating and processing the material.

Equisetales

Equisetum cones differentiate underground during the late summer and contain mature spores when they emerge from the ground the following spring. Young strobili should be dissected away from the rhizome, thoroughly washed, divided into several pieces, and killed in medium chrome-acetic or *FAA*. Both transverse and longitudinal sections should be made. There is little excuse for sectioning strobili containing mature spores. Compared with a dissection under a binocular, a section presents an utterly inadequate picture of the interesting organization of the cone. Mature spores should be studied in a wet mount, which is subsequently uncovered and permitted to dry, bringing about the uncoiling of the elaters.

Gametophytes can be grown by sowing newly shed spores on sterilized sphagnum. Excellent preserved gametophytes also can be purchased. Embedding and sectioning are carried out as with other soft, delicate subjects.

Ophioglossales

Botrychium is the easiest member of this order to use for the study of reproduction. The sporophylls can be teased apart and crushed to determine the stage of sporogenesis. Subdivide the fertile frond into small pieces, and kill in medium chrome-acetic or Craf II. A wide variety of stains will produce brilliant preparations.

Collectors find gametophytes to be extremely abundant in localized areas during favorable seasons. Preserved gametophytes can be purchased and are easy to process.

Filicales

The position and construction of the sporogenous area or sorus and the character of the sporophyll differ in the numerous genera. *Asplenium nidus-avis,* the bird's nest fern, bears sori on the large, leathery, entire vegetative leaves, whereas *Onoclea struthiopteris,* the ostrich fern, bears the sporangia in the tightly infolded, pod-like pinnules of special fertile fronds.

The preparation of the diverse subjects is practically identical. Select young sori, and examine a dissected portion of a sorus, using stages up to and including young thin-walled spores. Excise small portions of leaf tissue bearing sori, and kill in medium chrome-acetic or Craf II. Species having soft leaves are more economically dehydrated in alcohol or acetone, but butyl alcohol is advisable for the tougher types. Stain in iron hematoxylin to obtain the best nuclear details and in safranin-fast green for general use. Do not waste time embedding mature sporangia. The contents of the sporangium, the construction of the annulus, and the character of the wall of the mature spore are shown far better in a wet mount of fresh or preserved material. Some of the cultivated ferns have a high ratio of shriveled, undeveloped spores in the mature sporangium; sections of such material are disappointing.

Gametophytes of native ferns can be found in great abundance by an experienced collector. Such materials are useful for gross study, but the presence of soil particles among the rhizoids makes sectioning difficult and unsatisfactory. Gametophytes can be grown on nutrient agar cultures, or on porous clay flowerpots in a moist chamber. Remove a few gametophytes for examination at intervals, kill desirable specimens in medium chrome-acetic or Craf I, and prepare whole mounts (Chap. 10) or embed very carefully for microtome sections. Iron hematoxylin and safranin-fast green yield beautiful preparations. There is no need to section thalli bearing sporophytes, and permanent whole mounts are not so desirable as wet mounts that can be handled and viewed from all angles.

Gymnospermae

Members of the common genera of the Coniferales are well-known trees of great economic importance, and abundant material is easily available. The life history of the pine is probably the most widely used subject, therefore, the present discussion will be centered around reproduction in the pine. The reader should consult Chamberlain

(1935) for the morphology and seasonal sequence of the reproductive cycle in other genera and orders, and adapt the methods described here to other subjects.

Staminate cones of *Pinus* are differentiated during the season prior to the shedding of pollen. Cones can be dissected from buds and the stage of microsporogenesis ascertained by means of acetocarmine smears. Several species of *Pinus* undergo meiosis early in May, in the Chicago region. Killing fluids do not penetrate readily into large masses of highly resinous tissues. It is therefore necessary to subdivide all but the very smallest cones. Kill in *FAA* for general morphological studies and in a Nawaschin type, such as Craf II, for more critical details. Nuclei of microspores and mature pollen grains are stained adequately in hemalum-erythrosin. For the first gametophytic somatic mitosis, which takes place in the microspores before they are shed, use iron hematoxylin or safranin-fast green.

Preparations of the ovule history are much more difficult and time-consuming to make than the pollen history. The time of occurrence of interesting and important states varies with the species, the locality, and probably in a given locality in accordance with the weather conditions. In the Chicago area the megasporocyte of *Pinus laricio* is evident when the cones emerge from the bud. Fertilization has been found toward the end of June, while early embryo stages are obtainable during July (Chamberlain 1935).

The deep-seated megasporocyte is not reached readily by killing fluids, necessitating the use of vigorous fluids that produce distortion. The very young cones may be fixed entire in strong chrome-acetic, *FAA*, or *FAA*–bichloride of mercury. Such preparations are of interest principally to the student of developmental morphology. It may be preferable to cut away the young ovules from the sporophyll and strive to preserve the sporogenous and gametophytic features. Strong chrome-acetic seems to have given the best results for most students of this group. The Nawaschin modifications and Allen-Bouin modifications deserve further study.

Staining of ovulate structures is particularly difficult. Resinous materials in the cells tend to make the preparations unsightly, although the essential nuclei may be clearly differentiated. Safranin-fast green meets the requirements for all but research needs.

After the first few divisions of the zygote, microtome sections are no longer adequate for the study of embryology. The development of dissection methods has facilitated great progress in such studies.

A detailed discussion of the morphology and techniques applicable to other orders of gymnosperms is given by Johansen (1940).

Angiospermae

The angiosperms are usually the central feature of the study of reproduction in plants, representing the climax in the development of reproductive organs. Floral types and features of floral organs are studied best by dissection and whole mounts of fresh or preserved material.

THE FLOWER

Microtome sections are indispensable for the study of vasculation and histogenesis of floral organs. Each species is virtually a problem in itself; therefore, this discussion will be limited to the methods used for the successful preparation of a few useful subjects. Buds of lily and tulip are among the most satisfactory subjects for entire flower buds. The very young buds are large and easy to handle. Embedded buds can be accurately oriented for sectioning, and the parts are so large that elementary students can locate and recognize the parts on the slide. Lily buds are available over a considerable period, beginning with *Lilium umbellatum* and *L. elegans* in May, to *L. tigrinum* in August. Well-developed floral parts are shown in buds that are less than 5 mm. long (Fig. 15.4). Cut off at the base of the perianth, and remove successive slices from the tip until the tips of the anthers have been cut off. Drop into the killing fluid and pump vigorously. Fair fixation is obtained in *FAA*, but superior results are obtainable with Allen-Bouin II. Sectioning and staining are delightfully easy. Begin sectioning at the base of the flower, discard the ribbon until the sections include anthers and ovary, and discard the block when ovules are no longer present in the apical portion of the ovary.

Buds of tulip for entire sections of young flower buds are obtained from bulbs during late fall. Many varieties of Darwin tulips are in a suitable stage from mid-October to early November. Meiosis was found to occur in several Darwin varieties in October. Kill in Allen-Bouin II and carry through an alcohol-xylene, dioxan, or butanol series for entire young flower buds; for an older ovary follow the recommendations for the lily. Cut open the bulb, and dissect out the complete flower bud. Trim and kill as with lily.

Matthiola, the common garden stock or gillyflower, furnishes a suitable dicotyledonous flower for complete sections. Remove individual flowers, trim the end of the closed perianth, and kill in *FAA* for gross study or in the fluids recommended for lily. Flowers of tomato also are excellent for advanced workers.

THE ANTHER AND OVARY

Microsporogenesis can be studied satisfactorily in the lily. The structure of the anther and sporogenous tissues are also shown well (Fig. 15.2). Whether meiosis in the anther is demonstrated with microtome sections or smears depends on facilities for the production of enough slides for class use. Slides of adequate quality for elementary classes can be produced in quantity by sectioning (Fig. 15.3 *a*), but smears are far superior for critical details (Fig. 15.3 *b*). For elementary work, the essential and more obvious features of meiosis can be demonstrated with paraffin sections from a series of anthers beginning with anthers 2 mm. long up to anthers that are just beginning to show color. Ascertain the stage by means of acetocarmine smears and handle the successive age classes in separate bottles. This saves much time in locating desired stages for sectioning. Subdivide young premeiotic anthers transversely into pieces not over 2 mm. long (Fig. 15.1 *A, B*). The excellence of fixation is influenced by the degree of subdivision. Good fixation can be obtained by slicing anthers into disks less than 1 mm. thick while holding them under the killing fluid, Allen-Bouin II. This fluid preserves the sporocytes and meiotic chromosomes well enough for elementary teaching (Figs. 15.2, 15.3 *a*). The anther pieces cannot be cut much shorter than 2 to 3 mm. because the sporocytes are loose in the anther at this stage. The chromosomes are superbly stained by iron hematoxylin, gentian violet-iodine, and safranin-fast green.

The advanced worker who wishes to demonstrate the intimate structure of the chromosome during meiosis should explore the rapidly expanding literature on smear methods, select a species on which to work, and strive to perfect his technique until he can demonstrate the structures described by investigators of the subject (Fig. 15.3 *b*).

The dyad condition and second or equational division are of very short duration in lily, and will be found in material selected and prepared by the foregoing methods. The quartet (tetrad) and microspore stages are of long duration, present during the long period of expansion of the flower bud, until the anthers begin to color. For general purposes it is adequate to kill the entire anther; *FAA* yields surprisingly good results. Test each species by means of whole mounts before making a collection for this stage. Many cultivated lilies,

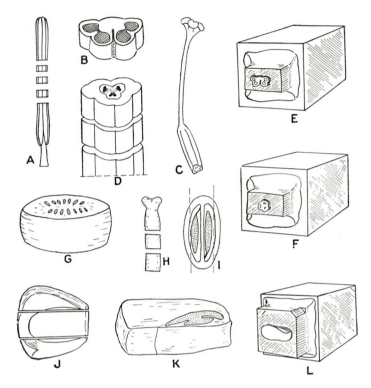

FIG. 15.1—Subdividing of reproductive organs: *A* and *B*, anther of lily; *C* and *D*, ovary of lily; *E* and *F*, mounted embedded blocks of anther and ovary, respectively; *G*, transverse disk sliced from young fruit of small-fruited variety of tomato; *H* and *I*, silique of *Matthiola*; *J*, kernel of corn sliced longitudinally; *K*, center piece of kernel containing essential parts of embryo; *L*, embedded kernel mounted for sectioning longitudinally. Trimmed edge of paraffin block produces a notched ribbon as in Fig. 6.4*A*.

especially the Easter lilies as well as *L. speciosum* and *L. umbellatum,* have extremely high pollen sterility, and the finished preparations show both nicely preserved pollen grains and shriveled microspores. However, such preparations are useful for illustrating pollen abortion. *Lilium regale, L. tenuifolium,* and *L. tigrinum* are particularly recommended for the study of pollen formation. The first two species have a high ratio of normal pollen, whereas only some strains of the last species are satisfactory.

For more critical fixation of microspore and pollen nuclei than is afforded by *FAA,* use the methods recommended for prophases. The somatic division of the microspore nucleus occurs over a brief period and is seldom encountered. The monoploid (haploid) chromosome complement is interesting and deserves careful staining when found.

Lily ovary is by far the most commonly used subject for teaching the development of the ovule and female gametophyte. The objection to lily is that the nuclear history of the embryo sac differs from the condition in corn, the legumes, and other common crop plants. However, lily ovary and its parts are large, the parallel seriation of the numerous ovules makes sectioning productive, and slides of the earlier stages, up to quartet formation, can be made economically in quantities (Figs. 11.4, 15.4, 15.5). Chrome-acetic has long been a favorite fluid for this subject, and formulas 0.5–0.5 and 0.3–0.70 are excellent for the smaller sporocytes (Fig. 11.3), but the results are rather uncertain with fully expanded sporocytes and subsequent stages. Bouin's solution has been used extensively, but the results are extremely variable. Figure 11.4 *b* shows a typical Bouin image that is all too common. The rims of the integuments often show a highly wrinkled and collapsed condition. The condition of the sporocyte and integuments after embedding can be determined accurately in a melted strip of paraffin ribbon. The proportions of ingredients in the original Bouin formula have been rather rigidly accepted by most users, but it is not improbable that superior results could be obtained with carefully determined variants of the formula. The author has obtained some excellent results by using propionic instead of acetic acid as suggested by Johansen (1940). The quality of the fixation is improved if the perfectly fresh ovaries are cut into thin disks.

The most consistent results for all stages have been obtained with the Allen-Bouin modifications, especially II and III (Table 3.2). A closely graded alcohol-xylene or acetone-xylene series can produce excellent results (Fig. 11.4 *d*), but failures are frequent. The glycerin-evaporation method, the dioxan series or *TBA* are much more reliable. The contents of the mature embryo sac are apparently highly fluid and particularly difficult to preserve without excessive plasmolysis, but Allen-Bouin II usually yields adequate fixation.

Staining sections of young ovaries prior to meiosis is one of the easiest tasks. A simple hemalum stain with or without erythrosin is adequate for elementary classwork. Iron hematoxylin and safranin-fast green yield brilliant preparations. The meiotic and gametophytic division figures and nuclei should be stained with iron hematoxylin, safranin-fast green or safranin-gentian violet. The last combination and the triple stain show the spindle fibers exceptionally well.

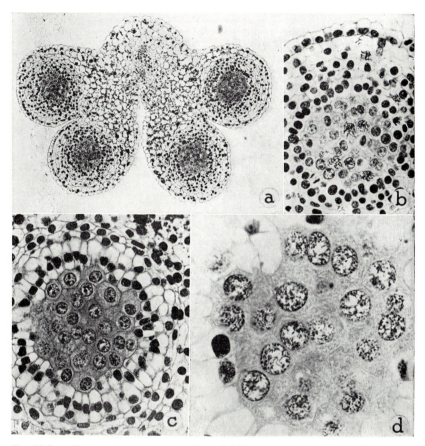

Fɪɢ. 15.2—*a,* Transverse section of anther of *Lilium regale; b,* somatic divisions in developing archesporium; *c,* archesporium, surrounded by differentiating tapetum; *d,* sporocytes in pre-leptotene phase.

The manufacture of lily ovary slides showing the seven-to-eight-nucleate stage is unproductive and expensive. Most of the slides obtained from a ribbon show incomplete embryo sacs. Cutting an ovule longitudinally through the center and having all the nuclei in one section is a matter of chance. Commercial manufacturers have a sales outlet for slides having incomplete sacs and can therefore sell the few choice slides having complete sacs at reasonable cost. For routine teaching, with its attendant breakage of slides, it may be more satisfactory to purchase slides of the seven-to-eight-nucleate stage than to make them. Good fixation has been obtained with fair regularity with Allen-Bouin II and *n*-butyl alcohol dehydration.

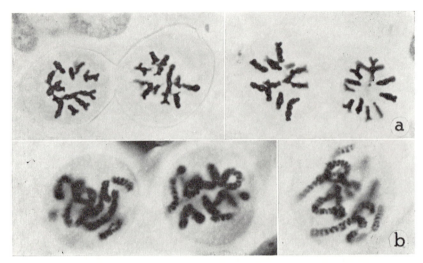

Fig. 15.3—*a, Lilium regale,* sectioned in paraffin, first division of meiosis in micro-sporocytes; *b,* smeared microsporocytes of *Tradescantia bracteata,* Sax-Humphrey method, iron hematoxylin.

Lilium represents a type of embryo-sac history that differs from the type found in many of our important crop plants. Slides of lily ovary are relatively easy and inexpensive to prepare. This plant should be used to show the transverse floral diagram of the flower bud (Fig. 17.4 *a*); the carpellary organization of the ovary (Fig.

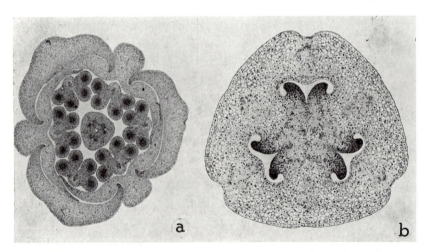

Fig. 15.4—*a,* Transverse section of flower bud of *Lilium regale; b,* ovary of same.

15.4 *b*); the origin and development of the ovule and integuments (Fig. 11.3); the origin and enlargement of the megasporocyte (Fig. 11.3); meiosis, (Fig. 11.4) and the four megaspores in the unpartitioned embryo sac (Fig. 15.5). It is of interest that the embryo sac of *Lilium pardalinum* is narrow and the megaspores are in linear order, whereas *L. umbellatum* has a broad embryo sac and a cruciate quartet (Fig. 15.5). The so-called normal type, which might better be named the common type, involves the formation of a quartet of megaspores, three of which degenerate, the fourth giving rise to the female gametophyte. This type occurs in maize, the legumes, tomato, and many other economic plants. The preparation of each of these is virtually a research task, and the reader who wishes to work on any of these plants should survey the literature on the desired plant.

Lilium is a good subject for making preparations showing fertilization (Fig. 15.6). Begin collecting 48 hr. after pollination and make collections every 12 hr. Use the killing fluids and methods recommended for the embryo sac. A series of collections will show stages from unfertilized mature embryo sac to young embryos (Fig. 15.6 *a*).

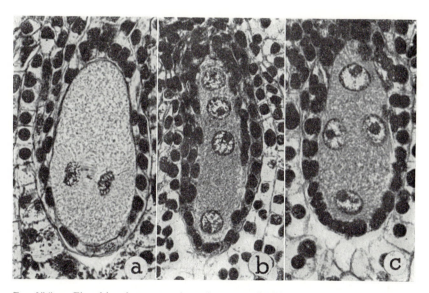

Fɪɢ. 15.5—*a*, First binucleate stage in embryo sac of *Lilium tigrinum.* Note the lagging chromosomes; *b*, linear megaspores of *L. pardalinum; c,* cruciate arrangement of megaspores of *L. umbellatum.*

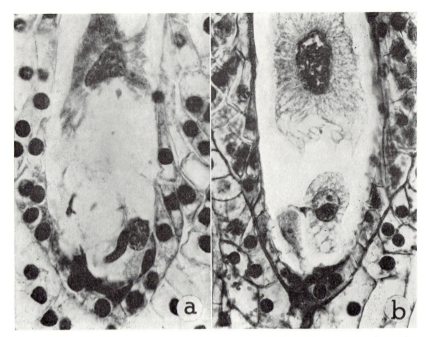

Fig. 15.6—Fertilization in *Lilium regale; a,* sperm and egg in contact; *b,* sperm appressed to egg, zygote wall evident.

THE EMBRYO, SEED, AND FRUIT

Embryology is usually neglected in elementary courses, in part because of the high cost of an adequate series of slides. Slides of early stages of embryo development are comparatively expensive to make, whereas the manufacture of slides of nearly mature embryos is more productive. The embryo of lily is large and not difficult to prepare. Use species that produce seed, such as *Lilium regale* or *L. tenuifolium.* Cut the ovaries into disks not over 2 mm. thick and divide longitudinally into three sectors, each sector containing one locule. Good fixation of the embryo can be obtained consistently with Allen-Bouin II, having the formaldehyde solution reduced to 5%. Section transversely at 15 to 18 μ. The flat developing seeds are in long tiers, and a block yields many good sections, all cut longitudinally with the axis of the embryo (Fig. 15.7). If six or eight sections are mounted on each slide, most of the slides will contain at least one accurately cut embryo. The most satisfactory stain is iron hematoxylin with a very light counterstain of fast green, which stains the cell walls of the embryo.

Slides of the caryopsis and embryo of maize are not difficult to make if the fundamentals outlined in the earlier chapters are observed. Consult the bulletins of agricultural colleges for the methods of making hand pollinations. Collect the ears at the desired intervals after pollination. Remove the husks carefully and trim away two rows of kernels without damaging the adjacent rows. With a thin, sharp scalpel cut off the intact kernels close to the cob and drop them into a Petri dish of water. Lay a kernel, with the germ upward, on a sheet of wet paper, remove chaff from the base and trim a longitudinal slice from each side of the kernel (Fig. 15.1 *J, K*). Also prepare some kernels for transverse sectioning by removing the basal and stylar portions of the kernel, saving only the portion from the tip of the coleoptile to the tip of the radicle. After the embryo is 25 days old, better infiltration of pieces for transverse sections can be obtained by transversely bisecting the embryo at the scutellar node, as well as removing the basal and stylar regions as above.

The essential morphological structures of the kernel are well developed in 25 to 30 days, and the pericarp becomes hard and brittle in 30 to 40 days. It is usually unnecessary to section the entire caryopsis after these dates. The embryo can be extracted easily between the 15th and 40th day, or until the kernel becomes so hard that the embryo is fractured if an attempt is made to dissect it out.

After the kernel has undergone maximum natural drying, or even if the kernel has been artificially dried for storage, the embryo can

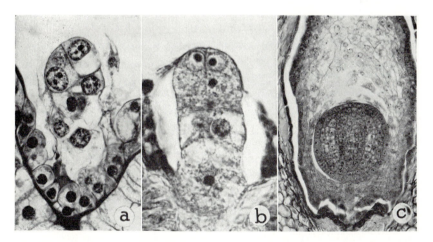

Fig. 15.7—*a*, Embryo of *Lilium regale; b,* embryo of *Lotus corniculatus* (courtesy of Dr. Harold W. Hansen) ; *c,* embryo of *Lycopersicum esculentum.*

be extracted. Soak the kernel in a solution that is based on the steeping liquor of corn processing plants. The solution used at present contains 2% sodium sulfite and 2% lactic acid. It is necessary to test the time and temperature factors with each lot of grain. Try 20° and 35°C., and intervals of 1 to 3 days. When the germ can be loosened easily, trim away unessential parts of the germ, subdivide if desired, and drop into the fixing fluid.

Small kernels, up to 10 days after pollination, are well fixed in Craf I or II. Older kernels and extracted embryos are penetrated better by Craf III. Xylene is the poorest solvent for infiltration, and chloroform is satisfactory only to the 15th day. Thereafter, a dioxan-normal or dioxan-tertiary butyl alcohol series makes possible the sectioning of 40-day kernels in paraffin (Sass 1945). (See frontispiece).

Embedded kernels must be soaked in warm water before sectioning. Adhesion of sections to the slide requires careful flattening of the ribbon, without overheating. For some research problems, hemalum alone permits adequate diagnostic observation. A safranin-fast green stain is attractive, and for exhibition purposes it is possible to make gaudy multiple stain preparations. Ages of kernels are given in the legends of figures 15.8 and 15.9.

Capsella bursa-pastoris is a favorite subject for embryology. The siliques are soft and easy to section. Although the seeds lie in the locules at various angles, seeds are so abundant that almost every section has complete embryos. Remove the fruits from the inflorescence, and classify them roughly into age groups in accordance with their distance from full-blown flowers. Process each class in a separate bottle. A sequence of stages in embryo development can be built up by sectioning fruits from the several lots. Trim two sides of the silique to promote penetration. The long silique of *Matthiola* also may be used. Divide transversely into 2- to 3-mm. lengths for killing, and cut microtome sections longitudinally. Use Craf I for either of these crucifers.

Lycopersicum esculentum, the tomato, is an excellent subject for dicot embryology. Use the small currant tomato, *L. pimpinellifolium*, seeds of which are obtainable from seed dealers. Slides of fertilization and the very young embryo are difficult and time-consuming

---

Fig. 15.8—*Zea mays*: *a*, kernel of pop corn 10 days after pollination; *b*, embryo of dent corn, 10 days; *c*, embryo, 15 days; *d*, transverse section of kernel of dent corn, 25 days; *e*, kernel of pop corn, 20 days.

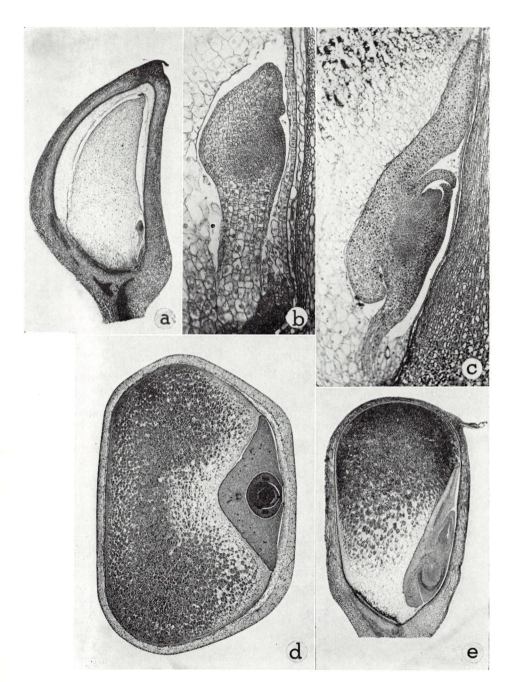

to obtain in quantity. Ten days after pollination the growing end of the embryo has developed into a large sphere that can be found in sections with adequate frequency (Fig. 15.7 *c*). Collect developing fruits of various sizes, slice out a transverse median disk of approximately one-third depth of the fruit, and kill in Craf I. Dehydrate and infiltrate with care. Section transversely, 10 μ for the earlier stages, 15 to 18 μ for larger embryos. As the seed coat of the maturing seed hardens, sectioning becomes increasingly difficult; however, the most important features of organogeny and the initiation of histogens are evident before the seed coat becomes excessively hard.

The legumes may be tempting subjects for embryology, but the best known large-seeded species, as well as many small-seeded species yield very few slides for the time expended. *Lotus corniculatus* is one of the most promising legumes. The long, straight ovary contains many ovules in straight alignment, and sectioning is fairly productive (Fig. 15.7 *b*).

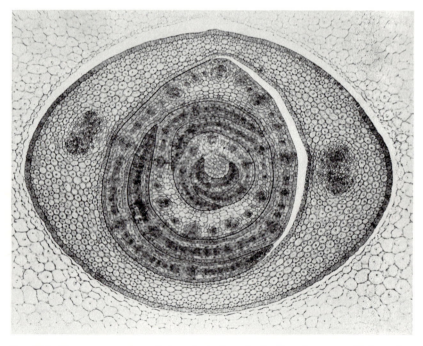

Fig. 15.9—Transverse section of plumule in kernel of yellow dent corn, 30 days after pollination (courtesy of Dr. Robert S. Fairchild).

Young fruits are processed in accordance with the methods given above for older ovaries. The currant tomato and many siliques are small enough even when nearly mature to have complete sections on a slide. Fruits that are more than 1 cm. in diameter should be subdivided and suitable pieces selected from the regions that are to be studied. The developing drupe of *Prunus virginiana* is an excellent subject. To prepare small cherry fruits for killing, remove a thin vertical slice from each side of some fruits and from the top and bottom of others, thus furnishing material for transverse as well as longitudinal sections. Kill in *FAA* for vascular study and in Allen-Bouin II or Craf III for better fixation of the embryo. The presence of brown pigmentation in many fruits produces poor contrast with the hematoxylins, but safranin-fast green is usually satisfactory.

The great array of fruits available to the technician presents a wide range of size, texture, and other properties, from the juicy berry to the flinty caryopsis. It is, therefore, quite impossible to offer generalized recommendations. The worker who ventures to prepare fruits and seeds has probably gained sufficient experience with easier subjects to adapt the fundamental methods given in this manual.

16. Microscope Construction, Use, and Care

The microscope is probably the most indispensable of the instruments used in the biological sciences. Intelligent purchase and effective utilization of a microscope require an understanding of at least the elements of its optical and mechanical construction. It is an expensive instrument, built to the highest standards of precision and having possibilities of performance that are not fully utilized by many users. Although having some structural features that are delicate or even fragile, the microscope has adequate durability to give many years of useful service.[1]

The function of a microscope is to produce an enlarged image of an object. This is accomplished by a system of lenses. A lens may be defined as a transparent body having at least one curved surface. A simple lens, consisting of one piece of glass, may be used to illustrate how a lens produces an enlarged image by bending or refracting light. A ray of light coming from the object enters the upper portion of the curved face of a lens and is bent downward. Similarly, a ray entering the lower portion of the lens is bent upward. The rays which pass through the lens converge and then continue as a diverging cone. If a sheet of paper or ground glass is placed to intercept the rays which pass through the lens, an enlarged image of the object is produced on the screen. A photographic plate can be placed in the cone of light and a photographic image obtained. A hand lens or the lens on a simple dissecting microscope produces an image on a screen in this manner (Fig. 16.1 *A*). The objective or lower lens of a microscope consists of two to nine lenses which act as a unit to produce an image as described above. There are certain limitations on the magnification and quality of image obtainable with the objective alone. The primary image produced by the objective is intercepted

---

[1] The author has drawn freely on the catalogues and service leaflets of the leading optical manufacturers.

and magnified, and improved in quality by an eyepiece. The eyepiece or ocular consists of two or more lenses working as a unit and having a fixed magnification. If a ground-glass screen, a sheet of paper, or a photographic plate is placed at any plane above the eyepoint of the ocular, an image is produced (Fig. 16.1 *B*). Note that the primary image is inverted and the projected image is erect.

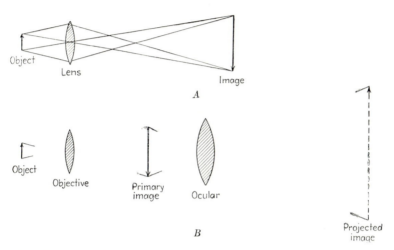

A

B

Fɪɢ. 16.1—Formation of projected images by the microscope: *A*, simple microscope; *B, compound microscope.*[2]

With a given objective and ocular, the size of the image varies with the distance of the screen from the ocular. If the screen is placed approximately 10 in. (254 mm.) from the eyepoint, the size of the image will be approximately equal to the product of the designated magnifications of the objective and ocular. Thus, an objective having a designated magnification of 10×, used with a 10× ocular, gives a total magnification of approximately 100×. Exact values must be determined by micrometry.

The foregoing discussion does not take into account the operation of the human eye working in conjunction with the microscope. However, most microscopic work is done by direct visual observation with the eye held at the eyepoint of the ocular. Let us turn for a moment to a consideration of the eye as an optical instrument. The lens of the eye operates as a simple lens, and the curved retina is

---

[2] The illustrations in this chapter are highly diagrammatic and simplified and are intended only to show the approximate relative positions of the object, the optical elements, and the images.

the receptive surface on which the image is formed. If an object is held at a given distance in front of the eye an inverted image of definite size is produced on the retina. If a larger object is substituted at the same distance, or if the original object is moved closer to the eye, the *visual angle,* or the angle of the cone of rays between the object and the eye, is increased, the size of the retinal image is increased, and the object appears to be larger. In Fig. 16.2A compare the two objects shown in solid and dotted lines, respectively, their respective visual angles Va_1, Va_2 and the retinal images Ri_1, Ri_2.

When the eye is held at the eyepoint of the microscope, it intercepts the image-forming cone which has a definite angle established by the microscope, and a retinal image of definite size is produced (Fig. 16.2 B). The observer sees a magnified virtual image, which appears to be near the level of the microscope stage, approximately 10 in. from the eye. The retinal image is erect, the virtual image is

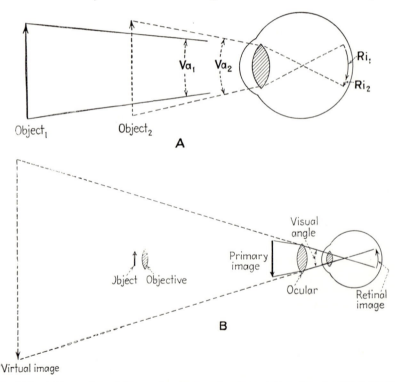

Fig. 16.2—A, Formation of images by the eye showing relative size of retinal image in relation to visual angle; B, retinal and virtual images obtained with a compound microscope.

inverted, and the direction of motion of the object is reversed. The apparent size of the virtual image is the same as if the observer viewed the projected image on a screen 10 in. from the ocular. As a specific case of magnification, let us view an object 0.1 mm. long with a 100× microscope. This produces a retinal image of the same size and the same impression of magnitude as if we looked at the 10-mm. image projected by the same microscope on a screen 254 mm. from the eyepiece.

Properties of Objectives

MAGNIFICATION

The most obvious property of objectives is magnification, which is a fixed value under the conditions outlined in a preceding paragraph. The objective magnifications used most commonly on standard monobjective microscopes range from 3.2× to 100×. Magnifications below this range are used on paired-objective stereoscopic prism binocular dissecting microscopes. Objectives above 100× have rather limited uses. The conventional low-power objective is 10×. The lower powers, from 3.2× to 6×, are not fully appreciated and deserve more serious consideration.

WORKING DISTANCE

Free working distance is the distance between the objective and the cover glass, using a cover glass 0.18 mm. in thickness. The catalogues of the manufacturers give the working distances of their objectives. A few selected illustrations show the relation between magnification and working distance: 10×, 7.0 mm.; 43×, 0.6 mm.; 45×, 0.3 mm.; 95×, 0.13 mm. It is obvious that for an elementary class the most desirable high-power objective has a magnification in the forties and the longest available working distance. Objectives of high magnification and short working distance must be used with care to avoid damaging the front lens and the slides.

FOCAL LENGTH

If a beam of parallel rays is projected through a simple lens, the rays converge at a point. The distance from this point to the optical center of the lens is the focal length. In an objective consisting of several components, the situation is somewhat more complex, and a different value is used. The manufacturers engrave on the mountings and list in their catalogues a value known as the equivalent focus. At the standard projection distance of 250 mm., an objective that has several components and an E. F. of 16 mm., will give an image of the same size as a simple lens of 16 mm. focal length. The E. F. should not be confused with working distance. The equivalent focus decreases as the magnification increases. The experienced worker is in the habit of speaking of an objective as a 4-mm. objective, for instance. For class use it is much better to speak in terms of magnification, which is 43$\times$ in a certain 4-mm. objective. In the past the manufacturers have paid undue attention to computing their objectives so that the equivalent focus is an even number, and a series of objectives will have the equivalent focus in the orderly progression 16, 8, 4, 2 mm., etc. A magnification may turn out to be some awkward fractional number like 3.2 or 5.1. A more practical series would be a sequence of magnifications such as 3, 5, 10, 20, 40, 60, etc. There is a trend toward the use of the latter system.

DEPTH OF FIELD (DEPTH OF FOCUS)

A minute body or a very thin section has thickness or depth. If a deep cell is being viewed with a 10$\times$ objective and the lens is focused on the upper wall of the cell, the bottom wall may also be in focus. If a 45$\times$ objective is focused on the top wall, the bottom wall may be completely out of focus and practically invisible. If the lens is focused on the bottom wall, the top wall becomes obscured. The vertical extent of the zone of sharp focus is the depth of field. The term "depth of focus" is in common use. For instance, some camera lens mounts have depth of focus scales. Depth decreases as the magnification increases. There are mathematical limits to the depth encompassed by a given objective. Magnification and other factors being equal, objectives of the several manufacturers have the same depth of field.

RESOLVING POWER

Resolving power is that property of a lens which makes possible the recognition, as distinctly separated bodies, of objects that are exceedingly close together, or subtended by a small visual angle. The simplest illustration of resolving power is the visibility of double stars. Although the two stars may be separated by a vast distance, the visual angle reaching the eye is very small, and the stars appear to be close together. Many individuals can see but one star. Other persons, whose eyes have better resolving power, can see the two stars distinctly. Applying this principle to the microscope, a lens of poor resolving power will show a slender chromosome as a single thread, whereas a lens of good resolving power will show the chromosome as two interwound threads. The question to ask concerning an objective is not "how small a thing can you see?" but "what is the minimal separation between two objects that the lens can resolve?"

The mathematical derivation of the formula for determining resolving power can be found in textbooks of optics or physics. The formula contains the following factors:

$n =$ the lowest index of refraction in the path of the rays, *i.e.,* the index of refraction of water, glass, air, cedar oil, balsam, etc.

$u =$ half of the angle made by the effective cone of rays entering the objective. This value can be obtained from a table in Gage (1941) or from the manufacturers.

N.A. $=$ numerical aperture, a number that is indicative of relative resolving power.

The formula is

$$\text{N.A.} = n \cdot \sin u$$

The value of the numerical aperture is engraved on most modern objectives and is given in the catalogues of the manufacturers. This number is 0.25 in a 10× objective, for example, and increases through progressively higher magnification, attaining the value 1.4 in an expensive 90× objective.

Knowing the numerical aperture, we can make a simple computation and arrive at a tangible value of resolving power. Assume that we are using an objective of N.A. 1.0 and using light having a wave length, in round numbers, of 0.0005 mm. The formula is

$$\frac{\lambda \ (= \text{wave length})}{2 \ \text{N.A.}} = \frac{0.0005}{2} = 0.00025 \text{ mm.}$$

This means that if two bacteria or two chromomeres on a chromosome are separated by a space of 0.00025 mm., they can be seen as two distinct bodies. As the numerical aperture increases, the resolving power increases, the working distance and depth of field decrease, and the cost increases.

The practicable upper limit of N.A. 0.95 is obtainable with dry lenses, used with an air space between the objective and the cover glass. In accordance with the foregoing formula, the N.A. can be increased by increasing the value of n or of *sin u* or both. If the angle of the ray that passes from the glass slide to air is greater than 41°, the light is totally reflected back into the glass. This phenomenon limits the angle that determines n. If a drop of cedar oil or synthetic silicone immersion fluid is used to connect the immersion objective with the slide, the values of *n* and *sin u* are both increased, and consequently the resolving power is increased. An N.A. of 1.10 can be obtained with a water immersion lens, and N.A. 1.40 using cedar oil.

OPTICAL CORRECTIONS

The foregoing discussion of the properties of objectives does not take into account the quality of the image produced. A simple lens produces a decidedly imperfect image. Rays of white light which enter the lens are broken up to some extent into a band of colors, a spectrum. These colors are not brought to a focus at a common point; blue is converged at a point closer to the lens than is red. Consequently, the colors of the object are not reproduced accurately, and a color fringe or "rainbow" is visible along the boundaries of objects in the microscopic image. This is known as chromatic aberration.

Spherical aberration is a defect that produces poor definition in the center of the field. This defect is aggravated by a cover glass that is not within the maximum thickness range of 0.15-0.21 mm. Image quality also is impaired by variation from the standard tube length of 160 or 170 mm. designated by the manufacturer. Certain objectives have an adjustable correction collar on the objective mount, calibrated for variations in cover-glass thickness.

Curvature of the field is another structural defect in the microscopic image. If the center of the field of view is in sharp focus, the margins may be out of focus. With some objectives, the image may be distinctly dome-shaped. The degree to which objectives are corrected for the above color and structural defects of the image will be indicated in the discussion of the optical categories in which objectives are classified.

PARFOCALIZATION

Two or more objectives are parfocal with each other when it is possible to focus one objective on an object, turn to the next objective without focusing, and see the object in more or less sharp focus. This feature is extremely important with large classes of elementary students. If the conventional 10✕ low power and the 40 to 45✕ high power are not parfocal, the student must refocus with the latter lens, which has a short working distance, small field, and shallow depth of field. Excessive breakage of slides and scratching of objectives occur if the objectives are not parfocal. Adjustment of the old-style objectives should be left to the manufacturers or to a skilled instrument mechanic. The new Bausch & Lomb objectives have an internal adjustment, with which the student cannot tamper but which can be easily adjusted with a special wrench.

Dry objectives between $10\times$ and $60\times$ can be made parfocal in any combination. The older 4 to $5\times$ objectives cannot be so adjusted, but the American Optical Company now makes a 3.5 and a $5.1\times$ objective, and Bausch & Lomb makes a 3.2 and a $6\times$ objective which can be made parfocal with the $10\times$, and parfocal with the $43\times$ within a quarter turn of the fine adjustment. With a combination of 3.2, 10, and $43\times$ objectives students should be taught to change magnification up or down in that sequence, thereby minimizing damage to slides and lenses.

Types of Objectives

ACHROMATIC OBJECTIVES

These are in the lowest price class and are used on classroom microscopes and for routine work in research. In these objectives chromatic aberration is corrected for two colors and spherical aberration for one color. Achromatic objectives have undergone relatively greater improvement in recent years than have other types.

FLUORITE OBJECTIVES

In these objectives the mineral fluorite is used in conjunction with special optical glasses. The corrections are of a higher order than those of the achromatic series. Fluorite objectives are particularly useful for photomicrography by virtue of excellent color correction. They are available only in magnifications over $40\times$.

APOCHROMATIC OBJECTIVES

These objectives have chromatic aberration corrected for three colors and spherical correction for two colors, affording brilliant images, presented in their true colors and without distortion of shape. The highly actinic violet rays are brought to the same focus as the longer visual rays, making these objectives highly desirable for photography. Apochromatic objectives are expensive because of their complex construction and the scarcity of suitable fluorite.

Oculars (Eyepieces)

Oculars have distinctive optical characteristics that must be understood in order to use the correct ocular, and the correct combination of ocular and objective under specific conditions. An ocular has a definite equivalent focal length. This value may be obtained from the catalogues, but a more useful designation, which is engraved on modern oculars, is the magnification value, which ranges from 4 to 30×. For routine work and for classwork 10× is the most satisfactory magnification. The lower magnifications are likely to have marked curvature of the image. Higher magnifications cause increasingly greater eyestrain, which is very pronounced with the 30×. Furthermore, there is an upper limit, beyond which the ocular produces only empty magnification, with no gain in the revealing of detail.

The maximum effective ocular magnification, which may be used with a given objective, can be computed easily. Assume that a 43× objective of N.A. 0.65 is being used; the formula is

$$\frac{1{,}000 \text{ (N.A. of objective)}}{\text{Magnification of objective}} = \frac{(1000) \ (0.65)}{43}$$

$$= 15\times, \text{ the approximate maximum ocular magnification.}$$

It is evident that with a microscope on which the 43× objective is the highest power used, an ocular magnification of over 15× is of no value with respect to resolving power but of possible value for counting or drawing by projection. This simple calculation will enable a purchaser to specify the most useful lens combinations. Modern oculars are parfocal, making it possible to interchange oculars of different magnifications without requiring much change of focus.

OPTICAL CATEGORIES OF OCULARS

Huygenian oculars are of comparatively simple two-lens construction. They are designed for use with achromatic objectives and yield inferior images with apochromatic objectives.

Compensating oculars are designed to correct certain residual aberrations inherent in apochromatic objectives. It is, therefore, imperative to use compensating oculars with apochromatic objectives, and oculars and objectives must be of the same make. These oculars may be used with achromatic and fluorite objectives having magnification over 40×.

Flat-field oculars are of the noncompensating type and yield images in which curvature has been considerably reduced. These oculars have various trade names, Hyperplane and Planascopic being the best-known. A serious objection to some oculars of this type is that the eye must be held rigidly at a restricted eye position. The slightest lateral motion of the head cuts off part of the field, and prolonged use produces marked fatigue.

Wide-field oculars (noncompensating) have an exceptionally wide field and good correction for curvature but may have a restricted eye position as in the flat-field type. This objection may be raised concerning high-eyepoint oculars, which are designed to permit the use of spectacles by the observer.

Workers who must use spectacles with low-eyepoint oculars find that the lenses of the spectacles and oculars become scratched after some use. A simple remedy is to paste a narrow ring of felt over each ocular. This permits the user to press his glasses against the ocular and to utilize the full field, without damage to the glasses or the ocular even after years of use.

Illumination

The most common method of illuminating a slide or other transparent object is by transmitted light. The light is projected through the hole in the stage and passes through the preparation. The simplest device for projecting light through the specimen is a concave mirror under the stage, designed to focus a converging cone of rays at the level of the specimen. Regardless of the character of the light source, whether daylight or artificial light, the *curved* mirror should be used if the microscope has no condensing lenses under the stage. The intensity of the illumination is controlled either by an iris diaphragm, or by a rotating disk having a series of holes of different sizes.

Microscopes that are used for advanced work are usually equipped with a condenser. A condenser is a system of two or more lenses under the stage, designed to receive a beam of parallel rays from a *flat* mirror or a prism and to converge the light at the level of the stage.

The simplest type of condenser, known as the Abbe condenser, consists of two lenses. Although Abbe condensers are not corrected for color or curvature, they are adequate for classwork and for much of the routine work in research. The N. A. is 1.20 or 1.25. The upper lens may be unscrewed (not in an elementary laboratory!); the lower lens then serves as a long focus condenser of N. A. 0.30, suitable for use with objectives of $10\times$ (N.A. 0.25) or less. On some Leitz models the upper element of the condenser is on a swinging yoke, whereby the upper lens can be swung aside, leaving the lower lens as a long focus condenser that fills the field of the lowest powers. A three-lens condenser with N.A. 1.40 is available for use with objectives having an

N.A. greater than 1.25. One or both upper lenses are removable, giving N.A. 0.70 and 0.40 respectively.

Aplanatic and achromatic condensers, made by several manufacturers, have excellent corrections for color and curvature. The elements, usually in 3 units, are separable, affording combinations with N.A. ranging from 0.20 to the full 1.30 or 1.40 of the complete condenser.

The resolving power inherent in an objective can be utilized only if the illuminating system has a numerical aperture equal to that of the objective. A curved mirror has an approximate N.A. of 0.25; therefore, it meets the aperture requirements of a $10\times$ (16 mm.) objective. Microscopes having objectives of over N.A. 0.25 should be equipped with a condenser, provided that the users are sufficiently skilled to use the condenser properly. An improperly adjusted condenser is worse than having no condenser. Some teachers prefer not to have condensers for large elementary classes in which thorough training in microscopy and close supervision are difficult.

The conventional high-power objective on elementary class microscopes is a 4–mm. objective, $43\times$ or $44\times$, N.A. 0.65 or 0.66. Many thousand instruments of this type are in use, equipped with an Abbe condenser of N.A. 1.20 or 1.25. If this condenser is not focused accurately it is a handicap, furthermore it does not cover the field of objectives below $10\times$. Removal of the condenser or of its upper element, a common practice among advanced workers when using low powers, is a most undesirable practice in large classes of beginners. The need for a condenser designed specifically for low and intermediate powers has been met by the American Optical Co. (Spencer Lens Co.) and the Bausch & Lomb Optical Co. These condensers have numerical apertures of .66 and .70 respectively, and therefore meet the aperture requirements of 4 mm., $43\times$ or $44\times$ objectives, and also illuminate the field of a $3.2\times$ or higher power objective.

A maximum N.A. of 1.00 can be obtained with a condenser if the condenser lens and the slide are separated by a layer of air. Obviously, an oil-immersion objective of N.A. 1.30 does not yield maximum performance unless the condenser, as well as the objective, is connected to the slide with cedar oil. Research workers who wish to obtain maximum resolving power make a routine practice of immersing the condenser. There are some practical objections to this practice for classwork.

Dark-field illumination is a neglected, but useful method of observation. In this method the light that reaches the eye from the object does not pass through the object but is reflected from the surface of the object. None of the light from the illuminant reaches the eye directly. The object thus appears to be self-luminous against a black background. Illumination of the object is obtained by either a standard condenser provided with an adapter or by means of a special dark-field condenser.

The simplest form of adapter consists of a wheel-shaped metal disk inserted into the slot below the condenser. The center of the disk cuts off the central rays of light and illuminates the object with the oblique marginal rays. A more effective adapter is a unit that replaces the upper element of the Abbe condenser.

The much more expensive dark-field condensers are of two principal types. Refracting condensers provide an oblique cone of light by refraction through the marginal regions of the condenser lenses. A disk below the central region of the condenser shuts out light from that portion. Reflecting condensers produce an oblique cone by total reflection from internal surfaces of the condenser lenses. Diagrams and descriptions of the various types of condensers can be found in the catalogues.

Dark-field illumination is recommended for the study of filamentous or unicellular algae and fungi, as well as for unstained sections of tissues. The cytoplasmic strands and nuclei of *Spirogyra* and cytoplasmic streaming in leaves of *Elodea* and filaments of *Rhizopus* make striking and instructive demonstrations.

The discussion of sources of light for the microscope has been deferred to this point, where the source can be discussed in conjunction with the condenser and the other optical components. Illumination is said to be *critical* when the source of light is superimposed on the object. This means that if an unfrosted tungsten coil bulb is the source, the coil is sharply defined upon the object. It is true that the portions of the object that coincide with the coil are under critical illumination, but only a very small part of the field may be so illuminated, and it is obvious that a naked coil cannot be used in this manner.

A frosted bulb is some improvement, but the granularity of the bulb is visible under critical conditions, as defined above, and the curvature of the bulb is visible under lower powers. If the condenser is lowered to obscure the granularity and curvature, the resolving power is decreased.

The desirable source is a flat, luminous, grainless surface of sufficient size to cover the field of the lowest power objective. When such a source is superimposed on the object field, uniform illumination is obtained. For elementary class use, and for many routine tasks in research, the nearest approach to optimum illumination seems to be an opal glass disk in a suitable lamp housing, with a 50 to 60 watt frosted blue mazda bulb. Place the lamp 8–12 inches from the microscope and manipulate the mirror until the field of view, with an object in focus, is uniformly illuminated. If the microscope has a condenser, place the point of a pencil against the lamp disk and adjust the height of the condenser until the pencil point is in focus. If finely ground glass, or a plano-convex lens with a ground flat surface is used instead of opal glass, the condenser must be moved out of optimum position to eliminate the granularity in the field of view. The use of a lamp that has a condensing lens system and a diaphragm is discussed in the chapter on photomicrography, and the worker who wishes to do critical visual work should consult that chapter.

Mechanical Operation

A microscope usually has a set of two to four objectives permanently installed on a revolving nosepeice. The objective are centered and parfocalized, each screwed into its designated opening in the nosepiece. The older nosepieces have adjustable stops for lateral centering of individual objectives. Improvements in manufacturing methods have made possible the quantity production of nosepieces of such precision that no adjustments for centering are required on the nosepiece. The removal of objectives should be strictly forbidden in the classroom.

The body tube of the microscope on which the objectives and ocular are mounted is moved up and down by two mechanisms, a coarse adjustment which produces rapid displacement, and a fine adjustment which moves the body tube very slowly. The coarse adjustment is actuated by a rack and pinion. This device is practically identical in the several leading makes. The tightness of the action can be adjusted easily by tightening or loosening the split bearing block against the pinion shaft by means of the readily accessible screws. In the Zeiss instrument the action is tightened by grasping the pinion heads firmly and screwing them toward each other.

The fine-adjustment mechanism differs radically in the different makes. One type employs a gear-and-sector device in which only a few teeth are in contact. This action, though very smooth and responsive, is rather delicate and easily damaged. The most rugged type is actuated by a split nut which has numerous threads in permanent contact with a worm gear. The threads are almost impossible to strip, and this action has excellent responsiveness. Details of construction of the various makes may be obtained from the illustrations and description in the catalogues. The repair of fine-adjustment actions should be entrusted only to a highly skilled mechanic or to the manufacturer.

The normal procedure in using the microscope is to locate the object with a low-power objective and then turn to the next higher power. Objectives of 10× or less are the most satisfactory finder lenses because of their large field of view, considerable depth of field, and long working distance. Microscopes for elementary work should be equipped with a safety stop on the body tube which prevents contact between the slide and the low-power lens. With an objective of 10× or less in position, it is safe to rack the body tube down until it is stopped by the safety stop. With the body tube in this position look into the ocular and manipulate the mirror until the field of view is uniformly illuminated. Move the body tube upward with the coarse adjustment until the image is visible, then bring the image into sharp focus with the fine adjustment. Search the section by moving the slide, using the fine adjustment freely to bring into sharp focus structural features at different depths in the specimen.

When it is necessary to turn to a higher magnification, center the desired structure in the field of view and bring it into sharp focus with the lowest power. *Without changing the focus,* turn the objective of *next higher* magnification into position. A properly parfocalized objective has ample clearance. The image should now be visible, and it should require not more than a quarter turn of the fine adjustment to bring the image into sharp focus.

The safety stop provided on the barrel does not prevent pressing the high-power objective upon the slide. Therefore, the high-power objective should never be used for locating the object. If an objective of 3 to 5× is used, do not change from this low magnification to 43×, but go progressively up through the range of magnifications. Similarly, go down the range progressively. The manufacturers can furnish safety stops for installation on the tubes of older microscopes.

Some teachers prefer to have the objectives adjusted so that when the object is located with the low power, and the high-power objective is swung into position, a slight *upward* movement brings the object into sharp focus. The objection to this arrangement is that, if the user inadvertently moves the body tube downward, he is moving it farther out of focus and may not stop until the slide is smashed. As an alternative arrangement the high-power objective may be parfocalized so that, when it is swung into position, the image is visible and a slight downward movement brings it into sharper focus. An accidental movement in the wrong direction, upward, will then do no harm. Students should be told firmly that there is no excuse for turning a fine-adjustment knob more than a half revolution in either direction. On the best modern microscopes very little pressure is exerted on the slide when the body tube is lowered upon it with the fine adjustment.

The condenser and illuminant are introduced again at this point. Assume that a grainless disk of a lamp serves as the immediate source of light. Adjust the height of the condenser until the surface of the disk is in view simultaneously with the focused specimen. The object is then within a disk of light of uniform luminosity. Obviously, the condenser does not project a point of light, but a disk of light in the plane of the specimen. Although a ground-glass source approaches the requirements for correct illumination, the condenser must be lowered to move the granularity out of focus. An opal glass disk permits a closer approach to correct illumination.

For general class use, the most practical light source is a lamp with a grainless or nearly grainless diffusing disk. It is preferable to have the lamp fastened to the table in constant relation to the position of the microscope. Under such conditions, the condenser, especially the low N.A. condenser described on page 192, can be mounted in simple ring mounts that are not adjustable by the students.

The position of the microscope in use depends to some extent on the height of the available table and chair in relation to the physical

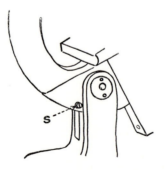

Fig. 16.3—Hinge stop for class-room microscope.

build of the user. Hard and fast rules of posture are ridiculous in a classroom having tables and chairs of fixed, uniform height, and students of diverse build. A very short person should certainly tilt the microscope for most work. However, if a fluid mount is used on a tilted stage, disturbing currents are likely to be set up in the liquid, and the liquid might drain into the diaphragm; therefore, it is advisable to use wet preparations on a horizontal stage. To forestall the progressive trend of weary students toward a reclining position, a hinge stop can be installed on modern microscopes, preventing tilting beyond 30° (Fig. 16.3).

Micrometry

The measurement of minute objects by means of the microscope is an interesting and valuable feature of microscopic study. Although the procedure is simple and rapid, the method does not receive adequate attention in teaching. The simplest form of measuring device is an eyepiece micrometer, a disk of glass having an engraved scale, a series of accurately spaced lines. The spaces do not have a standard value, and each disk must be calibrated for each given ocular and set of objectives. Place the disk upon the metal diaphragm in the ocular. If the diaphragm is in the correct position, the lines on the disk will be in sharp focus. Occasionally, these diaphragms become displaced, but they can be pushed back and forth with a softwood stick until the eyepiece micrometer is in focus.

The stage micrometer with which the calibration is made is a slide bearing an engraved scale with known values, usually in tenths and

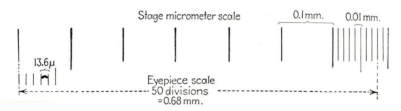

F1g. 16.4—Calibration of an eyepiece micrometer disk and measurement of a minute object.

hundredths of a millimeter, but scales in hundredths of an inch are obtainable. When the stage micrometer is brought into focus, the scale of the eyepiece will be seen superimposed on the scale of the stage micrometer. Shift the stage micrometer and revolve the ocular until the two scales are in such position that the values may be compared. A specific case using a 43× objective and a 10× ocular is shown in Fig. 16.4. It will be seen that the 50 small divisions of the ocular scale, only five of which are shown in Fig. 16.4, are equivalent to 6.8 large divisions or 68 small divisions of the stage micrometer scale. The computation is:

> 50 eyepiece divisions = 0.68 mm.
> 1 eyepiece division = 0.0136 mm. = 13.6 microns (μ)

The curved spore in Fig. 16.4, occupies one space on the ocular scale, and is 13.6 μ long.

The loose eyepiece disks described above are easily lost if they are not kept permanently in the ocular. In a large department it is an economy, over a period of years, to buy special micrometer eyepieces instead of disks. These eyepieces have a built-in disk, and the eye lens is adjustable to focus the scale sharply for the eyes of different individuals. Consult the catalogues for descriptions of micrometric devices.

Microprojection

The discussion of image formation showed that an image is produced if an intercepting screen is placed above the eyepoint of the ocular. With a sufficiently darkened room, a brilliant light source such as an arc lamp, and a good screen, an acceptable image can be obtained with the highest powers of the microscope. However, the most satisfactory results are usually at low and moderate magnifications. An image can be projected on drawing paper and a diagrammatic or detailed drawing made with considerable accuracy. Calibrations must be made for each lens combination and projection distance. This is done by projecting the image of a stage micrometer on the screen, measuring this image with an accurate ruler, and computing the magnification.

The catalogues and service leaflets of the manufacturers furnish detailed descriptions of a wide range of types and price classes of microprojectors.

Types of Microscopes

In the foregoing discussion of the elements of microscopy, the various types and makes of microscopes were not specifically discussed. A *simple* microscope is one that uses only one lens unit to magnify the object. The lens unit may be a single lens. A pair of lenses in fixed relation to each other comprise a *doublet*; a *triplet* consists of three lenses in a mounting. The most useful magnifications range from 6 to 12×. Magnifications up to 20× are available, but, as the magnification increases, the size of the field and the working distance decrease.

A *compound* microscope is one in which a lens unit, the objective, produces a magnified image, which is in turn magnified by a second lens unit, the ocular. By far the most common type of compound microscope employs one objective and one ocular in working position at one time. This is known as a *monocular monobjective* microscope. This type is durable, has a wide range of usefulness, and permits full use of the performance capacities of the optical system. The principal objection is that the user employs one eye at a time, and the tendency to use one eye more than the other causes excessive eyestrain and fatigue.

A *binocular monobjective* microscope uses a matched pair of oculars with a single objective. A system of prisms in the binocular body tube splits the beam coming from the objective and produces two images of identical magnification and intensity. The use of both eyes diminishes eyestrain and fatigue, and there is an impression of depth and perspective to the visual image. Ocular tubes of the binocular body are parallel in the majority of the principal makes. The tubes of American Optical Co. binoculars converge, but this firm will furnish parallel tubes. Convergent tubes present the image to the eye as if the image were at ordinary reading distance. When using parallel tubes the eyes are relaxed, as in looking at an object at a considerable distance. Some microscopists are convinced that they can use only one or the other of these two types of binocular with comfort, whereas other workers can use either type effectively. The binocular body has adjustments for separating the ocular tubes for the inter-pupillary distance of the observer. One ocular tube has a vertical adjustment for correcting slight differences of focus of the two eyes. To make this adjustment, select a minute structure in the specimen, close the eye over the adjustable tube and focus on the object with the fixed tube. Now close the eye over the fixed tube and bring the image into sharp focus in the adjustable tube with the focusing device on this tube.

The quality of the image obtained with binocular bodies is equal to that obtained with the single tube. Supplementary binocular bodies that are designed to be placed upon older monocular microscopes, have the tube length increased by the superimposed binocular body. A reducing lens system must therefore be used to bring the magnification back to the standard designated value. The most modern, and in many ways most desirable, binocular body has the eyepiece tubes inclined. This permits the head to be held in a comfortable position and greatly reduces fatigue.

An important category of binocular microscopes utilizes matched pairs of objectives. This type is customarily known as the dissecting binocular or *stereoscopic binocular*. These instruments show true perspective and depth. The image is erect, thus facilitating dissection, isolation, and other manipulations of the object. The practical range of total magnifications is from 10 to 150×. Two or more pairs of parfocal objectives can be installed on a nosepiece of either the revolving or sliding shuttle type. In one A. O. C. model a set of objectives may be permanently installed on the objective changer, a desirable arrangement for class use. For research work, each pair of objectives may be obtained in a removable mounting, readily inter-changeable on an objective changer, which, in the several makes is either a rotating drum, a rotating disk, or a sliding shuttle.

Several categories of noncompensating oculars are available for twin-objective binoculars. The standard Huygenian type is the least expensive and probably the most satisfactory for classwork. Wide-field oculars are well worth the greater cost. Two manufacturers produce a good junior-wide-field ocular, intermediate in cost and performance between Huygenian and wide-field oculars. High eye-point oculars also are available, but they require that the eyes must be held at restricted eye position, making these oculars objectionable to some workers.

This chapter would be incomplete without a few words concerning the durability and life span of the microscope. It must be obvious that the period of service obtainable from a well-constructed microscope depends upon the skill and care with which it is used, the amount of use, and certain environmental conditions, such as atmospheric conditions, extremes of temperature, and corrosive chemical fumes. An outstanding illustration of durability is afforded by an occasional microscope that seems to be in excellent mechanical and optical condition after 30 years of continuous research service. On the other hand, a classroom instrument may be in poor condition after 10 years of use. Serious scratching and corrosion become evident first on the 4-mm. dry objective, the oil-immersion objective, and on oculars, especially the type having a raised eye lens. The lower power objectives should show no contact wear or corrosion, especially if the instrument has a safety stop on the body tube. Examination of large numbers of class microscopes has shown that the serviceable period of a microscope is approximately 20 years. Replacement of the ocular and high-power objectives after 15 years is a good investment which may extend the life of the microscope for another 15 years. Periodic mechanical overhauling and refinishing of metal parts should be done by a competent fine-instrument mechanic. Major repairs and lens work should be entrusted only to the manufacturer. Considering the first investment, the low cost of upkeep, the large trade-in allowances, and the many generations of students served during a normal life span of a microscope, this instrument is the least expensive item of laboratory equipment.

The foregoing brief discussion of the principal types of microscopes and of the essential optical and mechanical features can be supplemented by a study of the well-illustrated descriptive catalogues of the leading manufacturers. Details of construction of specific models are available in leaflets provided by the manufacturers.

The belief in the superiority of the continental European optics may have been well founded 50 years ago, but is no longer a prime factor in purchasing an instrument. A choice among the better-known makes is now largely a matter of personal preference. The prospective purchaser should examine and, if possible, use various models and base his preference on mechanical and optical features and specifications that meet his needs.

17. Photomicrography

The use of photomicrographs for illustrations in teaching and research has become a firmly established practice. A choice between drawings and photomicrographs should be based on an understanding of the limitations and possibilities of these methods and upon the method of reproduction to be used. A drawing may be said to expound and explain the subject, while a good photograph is an accurate, impersonal reproduction of the subject. A drawing may be a routine diagrammatic record of rather gross structures, or it may represent the *interpretations* of the microscopist, either in full detail or in idealized, semidiagrammatic form. The routine type can be made by an artist; the interpretation drawing can be made only by the investigator sitting at his microscope. Photographs have similar characteristics and range from mere routine recording to the most critical probing of structural details.

Instead of arguing the relative merits of drawings and photographs, the experienced and versatile worker simply decides which method will best serve a specific need and uses such talent as he has or can hire. A few simple examples will illustrate the criteria by which a choice can be made between methods of scientific illustration. A cross section of a corn stem, or the corn kernel in the frontispiece contains several thousand cells. To make a drawing which would purport to be an accurate cell-for-cell representation would be an almost incredibly laborious task (for someone else to do). A photomicrograph of such subjects reproduces with acceptable accuracy the number, distribution, shapes, and sizes of the numerous cells and, furthermore, reproduces texture in a way that can only be remotely approached by the most talented artist. Photomicrographs of this type can be made only by a photographer who is familiar with plant materials.

Controversial subjects or new and striking discoveries deserve photographic illustrations. The reader has greater confidence in a description if it is evident that the investigator had presentable preparations. In illustrating some materials the very act of making an ink drawing on paper exaggerates magnitude, visibility of details, and texture. For instance, protoplasm does not consist of discrete dots and sharp lines. A photomicrograph accompanied by an interpretation drawing affords much more convincing illustrations of many subjects than does either method alone.

The making of record photomicrographs is often an essential part of diagnostic routine in clinical, chemical, criminological, and many other studies. Under standardized conditions, especially if there is some uniformity in the character of the subjects, such photomicrographs can be made by a well-trained technician.

In some fields of research the investigator is the only one who can locate and recognize the structures being studied. He must determine the proper focal level, the correct magnification, color filters, and other factors. It may be necessary to make several negatives at different foci in the same field of view. The investigator must personally decide from a contact print whether the photograph shows the desired structures. Research photomicrography of this type is clearly an inseparable part of the research and must be done by the investigator in person, with his research microscope and frequently without disturbing the slide that has been under scrutiny.

It is a common fallacy that a photomicrographer must be primarily a photographer, who can easily and quickly "pick up" what he needs to know about the microscope. On the contrary, he must be a skilled and critical microscopist, furthermore he must be familiar with the structure of the material that is to be photographed. He can learn the processing of negatives much quicker than he can gain a mastery of microscopy. Given a good negative and some supervision by the scientist, the commercial photographer can make excellent contact prints and enlargements.

This chapter was written for the research worker or teacher who has modest facilities for making photomicrographs and wishes to utilize them to the best advantage. It will be assumed that the advanced worker who has more elaborate facilities has studied both photography and microscopy beyond the elementary scope of this manual.

Attachment Cameras

Photomicrography can be done with a standard microscope and a camera attachment that rides on the ocular tube without other support. The largest camera of this type uses 9- by 12-cm. or $3\frac{1}{4}$- by $4\frac{1}{2}$-inch plates or films. The projection distance is constant, and magnification is equal to the product of the magnifications of the objective and ocular. Other models use smaller plates or sheet film, or 35-mm. or Bantam roll film. Examples of such attachments are the Erb and Gray Visicam, Leitz Makam and Micro-Ibso, Zeiss-Winkel, and the Bausch & Lomb Model N (Fig. 17.1). Precise focusing is possible with the lateral observation ocular or screen. If the collar that fastens the camera assembly to the ocular tube of the microscope does not hold the assembly rigidly, press or solder a brass sleeve to the outside of the ocular tube, and turn the sleeve on a lathe until it makes a tight, turning fit inside the clamping collar of the camera.

The Leitz, Zeiss, and similar attachments are advertised to be used with expensive miniature cameras like the Leica or Contax on the grounds that the camera can be used for "other purposes." This can be a handicap, rather than a virtue, especially if the camera must be shared by several workers. When one person wishes to use the camera for photomicrography, it is sure to be out on a field trip or a vacation trip, or it is loaded with the wrong film. The same criticism applies to the Exakta and similar reflex cameras. A stripped camera body[1] that is merely a holder for a spool of film can be purchased from Brinkmann,[2] Kessel,[3] and other dealers for use with the Leitz Micro-Ibso or the Zeiss-Winkel attachments (Fig. 17.1). By having several of these inexpensive film holders, several emulsions can be kept available.

An obsolete model, Exakta or similar reflex camera, stripped of lenses and other superfluous parts, can be bought for a small sum. Fastened permanently to a swinging bracket like the one in Fig. 17.3, the camera is always at hand. If the reflex device registers accurately with the emulsion plane, and if the shutter operates without jarring, excellent photomicrographs can be made at all magnifications.

---

[1] Based on the Argus, Eastman Pony, Bower, and similar cameras.
[2] Brinkmann Instrument Co., 115 Cutter Mill Rd., Great Neck, L. I., N. Y.
[3] W. H. Kessel & Co., 510 N. Dearborn St., Chicago, Ill.

Fig. 17.1 — Cameras that ride on the ocular tube: upper left, Leitz Makam 9- by 12-cm. fixed-length camera; upper right, Leitz 35-mm. Micro-Ibso (=Mikas), with simple film holder; right center, Viscam 35-mm. camera with Argus body; lower left, Bausch & Lomb 2¼- by 3¼-inch camera; lower right, B & L with 35-mm. film holder, interchangeable with 2¼- by 3¼-inch box.

Pillar-Type Cameras

This type of apparatus carries the camera on a vertical post, which is attached to a heavy metal base. A simple version has a bellows camera, which my be used either with a compound microscope (Fig. 17.2), or with Micro Tessar lenses that are carried in a focusing mount on the shutter. Excellent work can be done with such an apparatus if the components, including the separate light source are correctly and rigidly aligned. The microscope cannot be used conveniently for visual study because the camera front must be raised and the pillar swung away from the ocular. In order to do both visual and photographic work, much time is spent in dismantling and reassembling the apparatus.

The swing-out type of camera permits free use of the microscope for visual work, and the camera can be swung into position accurately. The lateral observation tube permits precise focusing (Fig. 17.2). A compact and rigid unit can be made by bolting a length of channel

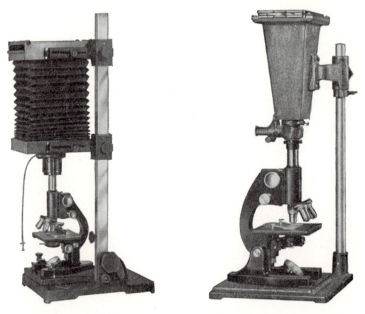

Fig. 17.2—Pillar-supported cameras: *left,* Bausch & Lomb bellows camera on hinged pillar; *right,* Bausch & Lomb fixed-length camera with observation eyepiece. Camera swings on post.

Fig. 17.3—Combination visual and photomicrographic apparatus, permanently assembled on commercial (B & L) metal base. The custom-built plywood camera swings on a post and can be removed.

iron to the metal base of either style shown in Fig. 17.2, and fastening the lamp permanently to the channel iron.

Combined Visual-Photographic Apparatus

Experienced research workers know that the taking of a photomicrograph is inseparably associated with critical visual study. For example, let us assume that a wet acetocarmine preparation has been studied with a fine binocular research microscope, and it becomes desirable to photograph a loose floating cell that is in satisfactory orientation. It is impossible to remove the slide from immersion contact with the objective and condenser, transport the slide to another part of the laboratory or another part of the city, set up the slide on another microscope, and locate the specific cell and photograph it in the original condition. Even if a permanent slide is used, the investigator — who is the only one who knows what is wanted — must personally locate the desired field and focus at the desired level.

These conditions call for equipment that permits a quick change from critical visual work to photomicrography, right in the research laboratory. Use a commercial metal base on which the microscope and the lamp are permanently fastened and aligned (Fig. 17.3). This arrangement, which can be used with attachment cameras that ride on the microscope, as well as with a pillar camera, permits comfortable visual study and a quick change to photography. The camera can be removed and used by another worker who has a similar base, microscope, and lamp. The versatile apparatus in Fig. 17.4 permits quick change-over from visual study to a 35-mm., Bantam, Polaroid, or 4- by 5-inch camera. The massive, rigid Bausch & Lomb reflex

FIG. 17.4 — American Optical Co. apparatus, convertible from 35 mm. or Bantam (left), to 4 by 5 inch (upper right), to Polaroid (lower right).

model L has the above desirable features, as well as convertibility to Micro Tessar and gross photography (Fig. 17.5). The microscope and lamp are on a base that is slid as a unit from under the camera for visual study. This apparatus could be improved by providing a separate base to which the removable microscope-lamp unit could be transferred so that someone else could use the pillar-base, pillar, and camera.

The remarkable Polaroid Land camera will greatly influence and perhaps revolutionize photomicrography. Used on the Central Scientific Co. assembly (Fig. 17.6) or the versatile American Optical Co. apparatus (Fig. 17.4), this camera produces black-and-white transparencies in minutes. From the transparencies, positive prints can be made for dissertations or publication. It remains to be seen whether such prints can compete with contact prints made from large, fine-grain negatives for producing half-tone cuts.

Optical-Bench Cameras

The optical-bench photomicrographic apparatus has long been considered the ultimate in precision and rigidity. The three principal components, the camera, the microscope, and the arc lamp, are

FIG. 17.5—Bausch & Lomb reflex camera that permits maximum versatility for visual study and photography, with Micro Tessar lenses as well as with all powers of the compound microscope.

Fɪɢ. 17.6 — Central Scientific Co. apparatus, with focusing device over ocular (left) and Polaroid camera swung into position over ocular (right).

mounted on a heavy metal track on which the units may be slid back and forth in accurate alignment. If the microscope is removed for visual study, the replacement and re-alignment are very time-consuming. There is a temptation to keep an expensive microscope, possibly a binocular, permanently on the apparatus where it is not available for visual use, and may be used for photomicrography only a fraction of the time (Fig. 17.7). For low magnifications, Micro Tessars are used in conjunction with special substage condensers. This requires removal of the microscope assembly and the installation and alignment of an entirely different optical system.

The sequence of operations for setting up and using these elaborate outfits is identical in principle with the procedures outlined for simpler apparatus. A lateral observation and focusing tube is available in some makes, or a ground-glass screen may be used for focusing. A limitation of present models is that they use large and expensive plates, 5- by 7-inch or 8- by 10-inch sizes. Reducing kits make possible the use of 3¼- by 4¼- inch and 4- by 5-inch plates. Continued improvement of fine-grain film of high resolving power will undoubtedly lead to the use of much smaller negatives, especially for expensive color work. The rapid trend to smaller cameras that can be used at the research bench in conjunction with visual study will probably reduce optical bench cameras to a minor role in photomicrography.

There is a continuing trend toward "trinocular" microscope heads, an inclined binocular to which a third, vertical tube has been added. The latter can carry a permanently attached 35-mm. camera, or a post-supported large camera can be positioned over the third tube.

In the ideal photomicrographic camera, nothing intervenes between the ocular and the emulsion during exposure.

Light Sources

The character of the light source and the method of illuminating the object are important factors in photomicrography. Artificial light is in almost universal use because of its constant intensity and ease of control. A 6-volt, 108-watt coil filament or ribbon-filament lamp furnishes a steady, fixed source of adequate intensity. A transformer furnishes 6-volt current from the 110-volt alternating-current line. A rheostat may be used to control the intensity if color temperature is not critical. The tungsten-arc and zirconium arc also are excellent illuminants. The carbon arc has a brilliant, homogeneous crater, but the crater shifts as the carbons burn away, and it is difficult to keep the crater exactly in the optical axis.

The lamp must be provided with an adjustable condenser and an iris diaphragm. A one-lens spherical condenser or the slightly more expensive aspheric condenser will give good results, but a better corrected condenser with two or more components is preferred.

Focusing Aids

The focusing panel in commercial apparatus is usually made with sufficient precision to place the ground glass in the same plane as the photographic emulsion. If correct register is not obtained, and

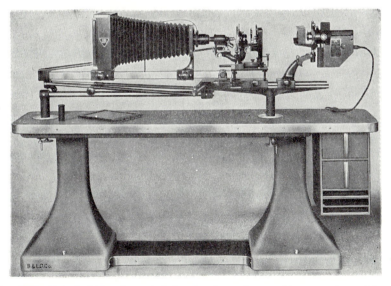

Fig. 17.7 — Bausch & Lomb optical bench photomicrographic apparatus with research microscope in place.

inaccurate positioning of ground glass is suspected, the easiest remedy is the use of a plate holder as a focusing panel. Remove the partition that separates the plates in a double holder. Insert a ground glass into the plate grooves. This places the ground surface in the same plane as the emulsion. Take the photographs with a plate or film holder of the same make as the one used as a focusing panel.

The ground-glass surface provides a satisfactory image for orienting the subject, but not for critical focusing. For maximum sharpness, use the clear window method. Make a fine X mark on the ground glass with India ink, on the diagonals of the glass. Allow the ink to dry, place a drop of balsam or cover glass resin on the mark and lower a cover glass on the resin. This will make a clear window in the ground surface. A focusing glass may be purchased, but an inexpensive one can be made by fitting a 3 to 5× magnifier into a metal tube of such length that when the tube rests on the clear area of the ground glass, the X mark is in focus. Bring the image into approximate focus on the ground surface, view the image through the magnifier and bring into sharp focus.

Exposure Meters

An extensive literature has accumulated on the subject of exposure control in photomicrography. The most accessible sources are the indexes of Stain Technology and the *Journal* of the Biological Photographic Association. Only a brief survey of the principal methods can be given here.

The photoelectric meters used for general photography will register a significant reading with some photomicrographic apparatus. This makes possible the calibration of the apparatus and fairly satisfactory exposure control. Several highly sensitive, but expensive, electronic meters are available. These meters give good readings in the plane of the emulsion, in fact permit probing of small areas of the image. Consult the advertisements of scientific journals for the currently available meters.

If an exposure meter is not available, an experienced photographer who can judge negative densities can obtain good negatives with a little expenditure of film and time. Assume that previous experience with a certain magnification suggests an exposure of 15 seconds. Draw the dark slide halfway out of the film holder and make a 10-second exposure. Remove the dark slide and expose for another 10 seconds. The two halves of the film have had 10 and 20 seconds respectively. Develop the negative and decide whether the next exposure must be less than 10 seconds, more than 20 seconds, or an intermediate interval.

Negative Materials

Orthochromatic emulsions can be used for photomicrography. These emulsions are sensitive to green, blue, and ultraviolet. A black object or one that is rich in green or blue may be rendered accurately in monochrome with such emulsions. Representative emulsions in this category and Eastman's ortho, process ortho and Verichrome films, D. C. Ortho plates, and Agfa Plenachrome film.

Noncolor sensitive emulsions such as process plates have not been given adequate attention for photographing such objects as black-stained chromosomes.

Historically, the best-known emulsion for photomicrography is the Wratten M plate. This is a panchromatic plate having comparatively coarse grain and slow speed, producing negatives of high contrast. The more recent fine-grain panchromatic emulsions may well bring about a radical revision of photomicrographic techniques. These emulsions are fast, they have a wide range of color sensitivity, and, because of the fine grain of the negative, enlargements of many diameters can be made. This makes possible the use of relatively low microscope magnifications, with greater depth of field and a large, comparatively flat field. A negative will yield a contact print or lantern slide of a large field, and selected portions of the negative may be greatly enlarged to exhibit finer details of structure. Films in this category are Eastman Panatomic X and Agfa Finopan. The speed ratings of emulsions can be obtained from the manufacturers or from the frequently revised tables of makers of exposure meters.

The choice between plates and films depends on the size of film, the microscope magnification being used, the type of negative holder, and the focusing method. Large sheet film has considerable concavity, whereas a glass plate is flat over its entire area. With the lower magnification ranges, up to 100$\times$, the lack of perfect flatness of the emulsion does not seriously influence focusing, but, if much of the area of a large negative is to be utilized with high magnifications, the use of plates may be necessary. Sheet film holders designed to hold the film along all four sides are superior to separate adapters that fit into plateholders. Some of these adapters do not hold the emulsion of the film in the same plane as when a plate is used in the same holder, therefore the focusing screen or observation tube is not in accurate register with the emulsion, resulting in inaccurate focusing. The foregoing sources of error should be tested for the available apparatus and accessories.

Roll film is useful only if the conditions are so well standardized that the length of exposure can be estimated accurately. The smaller sizes lie sufficiently flat for moderate magnifications. Pack film has some advantages over roll film. Individual films can be removed from the pack for development, making it possible to establish exposure time with one or more trial exposures, developing the films at once. Subsequent exposures under similar conditions can then be made in rapid succession. In the larger film-pack sizes the film has considerable curvature along the edges, but the central portions are adequately flat.

The Setting Up and Operation of the Apparatus

Before outlining the procedure used in taking photomicrographs, some suggestions are offered concerning the choice of objectives and oculars for any given subject. The ultimate aim of the photographer is a finished print on paper, or a lantern-slide (transparency) image on a screen. The image should convey to the observer the intent of the photographer: a low-power survey of a large area, with little emphasis on cell detail; a rendering of texture and tone in black and white, without much cell detail; an accurate reproduction of details within a cell or within a minute object; or the sharp outlining of an object against a contrasting background, without detail within the object. The worker may have other aims and may combine them, with emphasis placed where needed.

When using the standard oculars that are used for visual work, the best results are obtained with oculars of moderate magnification, 8 to 12×. The major manufacturers advertise special photographic oculars that produce a flat field. These oculars cannot be used for visual study.

The objective to use is one that covers the desired area of the object generously, especially when using visual oculars, so that the important area will be in focus simultaneously and the out-of-focus marginal region can be masked out in the finished product. In addition to adequate coverage, the objective should have adequate resolving power to show the *necessary* detail. Keep in mind that, as the magnification and resolving power increase, depth of field decreases. It may be advantageous to obtain a sharp negative covering the necessary area and depth of the object — but having relatively low magnification — and to enlarge a few diameters in making the positive. However, the positive must show the detail that the photographer intended to show. Some workers prefer to keep the negative image of such size that lantern slides may be made by contact, or that contact prints will be of the correct size for publication in a journal. Wider use of the fine-grain methods of miniature photography will promote the use of excellent objectives of comparatively low magnification, large field, and good resolving power. Examples are the Bausch & Lomb oil immersion, 40×, N.A. 1.00, and several makes of oil-immersion objectives, with magnifications of 60 to 65×, N.A. from 1.30 to 1.40.

The sequence of operations leading up to making the exposure will now be described. It will be assumed that the slide, all lenses, the mirror, and the filters are perfectly clean, and that all units are firmly fastened in place. The procedure varies with the type of illumination being used.

When using an ordinary mazda bulb behind a sheet of ground glass or grainless opal glass the operations are as follows:

1. Locate and focus the object as in visual study.

2. Place a thin wedge of black paper against the diffusion glass of the lamp, and focus the condenser until the paper marker is in focus simultaneously with the specimen. Remove the marker. If ground glass is used, the grain of the glass will be visible, and the condenser must be displaced slightly to eliminate this grain.

3. Remove the ocular and adjust the substage diaphragm until the back of the objective is just filled with light.

4. Replace the ocular, bring the camera into position, and adjust the angle of the mirror until the illumination on the focusing screen is centered.

5. Focus the image sharply on the focusing screen and make the exposure.

The use of the foregoing equipment and procedure may be regarded as amateur photomicrography, which nevertheless affords valuable training and may yield results that meet some needs.

For serious and critical work, the lamp should have a concentrated filament bulb, a condenser system of one or more components, and an iris diaphragm.

Two systems of illumination are possible with suitable lamps. *Critical* illumination is obtained when the condenser system focuses the incandescent light source (filament) upon the plane of the specimen on the stage. This superimposed filament image must be of adequate area to cover the field of the objective and must be of uniform brilliance. Many laboratories do not have a lamp suitable for this system, and it is not used extensively.

The *Köhler* system of illumination is the most practical and widely used method. The image of the filament is focused on the substage condenser diaphragm, and the image of the lamp diaphragm is focused in the plane of the specimen. The operations usually are performed in the following order:

1. Direct the beam of light upon the mirror, with no filters or other screens in the beam. Open the substage diaphragm completely, reduce the lamp diaphragm aperture, and manipulate the mirror until the light reflected back from the lower lens of the substage condenser is projected by the mirror as a spot of light, exactly centered on the lamp diaphragm. This position of the mirror must not be altered. If the filter holder is adjustable, insert any dense filter and adjust the holder until the light that is reflected from the back surface of the filter is centered on the lamp diaphragm.

2. Open the lamp diaphragm, close the substage diaphragm and focus the lamp condenser until the filament image is sharply defined on the substage diaphragm.

3. Bring the object into focus with the objective that is to be used to take the photograph.

4. Open the substage diaphragm completely and partly close the lamp diaphragm. Rack the substage condenser up and down until the lamp diaphragm, with its edges sharply defined, is superimposed on the sharply focused specimen.

5. Replace the ocular with a pinhole ocular, look down into the tube and bring the spot of light in the back lens of the objective

into the exact center by manipulating the centering screws of the substage condenser. (Not by moving the mirror.)

6. Replace the ocular. Open the substage diaphragm fully. If the disk of illumination — which represents the lamp diaphragm — does not cover the desired area of the object, remove the upper element of the substage condenser and repeat operation 4. It may be necessary to remove the upper two elements of a three-element substage condenser to obtain a large enough illuminated field.

7. Place a 3× to 6× magnifier above the ocular, adjust the magnifier until the back lens of the objective is in focus. Close the substage diaphragm and open it slowly until the rim of the diaphragm coincides with the rim of the back lens of the objective. The full numerical aperture of the objective is utilized only under these conditions. In practice, the aperture may be reduced by means of the substage diaphragm, but not more than one-sixth of the diameter of the back lens of the objective.

Up to this point the operations are identical for visual study and photography, and the foregoing operations can be performed with the binocular body. This is the time to try Wratten filters — usually in pairs — to obtain the desired contrast or detail.

8. Connect the camera with the microscope. Arrange the composition of the image on the screen by means of the revolving stage or revolving camera. Focus critically with a magnifier over the clear window of the screen. Close the shutter, insert the film holder, and make the exposure. With cameras that have a side-ocular, composition and focusing can be done after the dark-slide has been withdrawn from the film holder. The prism of the side-ocular can be swung aside and the exposure made.

Low Power Photomicrography With Micro Tessar-Type Objectives

The lowest power objectives, such as the 3.2× and 4×, do not cover a large enough field, and have too much magnification for some subjects. The objectives of stereoscopic binocular microscopes have the desired specifications, but such objectives are not adapted for use on a single-tube microscope of correct tube length. The compound system is therefore not suitable for photomicrographs in the 5× to 30× range (See frontispiece). Such photographs are taken with special objectives that produce a flat, well corrected image and are never used with an ocular. Objectives of this type are the Micro Tessars of Bausch & Lomb and the Micro-Teleplats of Spencer (American Optical Co.). Leitz and Zeiss also make excellent objectives of this type. Each Micro Tessar must be used with a condenser that has the same focal length as the objective. The manufacturers furnish matching condensers for their objectives. The illuminant must have a flat ribbon filament or a homogeneous arc, and a condenser lens.

The mechanical set-up and the operation of a particular commercial apparatus should be obtained from the directions supplied by the manufacturer. The principles will be outlined on the basis of the apparatus shown in Fig. 17.8. The usual sequence of operations is as follows:

1. Adjust the lamp condenser to give a beam of parallel rays. This can be done with adequate practical accuracy by focusing the filament upon a wall 8–10 ft. away. Lock the lamp condenser.

2. Remove all optical components from the stage and camera and center the filament image on the ground glass of the camera back. The filament will not be in sharp focus, but do not change the setting of the lamp condenser.

3. Insert the Micro Tessar objective and center it into the filament image by shifting the objective (lens board), not the light beam.

Fig. 17.8 — Horizontal apparatus for use with Micro Tessar objectives. Components, from left to right: ribbon filament lamp; filter holder; revolving stage with condenser holder; bellows camera with removable lens board, which carries the focusing mount and a behind-the-lens shutter. The commercial spring-back focusing screen is removable.

4. Insert the substage condenser that has the same focal length as the objective. Center the condenser into the beam by moving the condenser, not the light beam.

5. Place the slide on the stage and bring the object into sharp focus on the ground glass.

6. Try various Wratten filters and use the combination that gives the desired balance between contrast and detail.

7. Make the exposure.

The above optical system transmits enough light, if the filters are removed, to register adequately on a good exposure meter. Exposure factors can be worked out for a given optical system, filter combination and type of negative material.

Bibliography

BEATTY, A. V.
 1937. A method for growing and for making permanent slides of pollen tubes. *Stain Tech.* 12:13–14.
BECK, C.
 1938. The microscope. Beck. London.
BELLING, JOHN
 1930. The use of the microscope. McGraw-Hill. New York.
BOWEN, C. C.
 1956. Freezing by liquid carbon dioxide in making slides permanent. *Stain Tech.* 31:87–90.
BUTTERFIELD, JOHN V.
 1937. An illuminating system for large transparent sections in photomicrography. *Jour. Bio. Photo. Assoc.* 6:155–161.
CHAMBERLAIN, C. J.
 1932. Methods in plant histology. 5th ed. University of Chicago Press. Chicago.
———.
 1935. Gymnosperms — Structure and evolution. University of Chicago Press. Chicago.
CHOWDHURY, K. A.
 1934. An improved method of softening hard woody tissues in hydrofluoric acid under pressure. *Ann. Bot.* 48:308–310.
CONGER, ALAN D., AND LUCILE M. FAIRCHILD
 1953. A quick-freeze method for making smear slides permanent. *Stain Tech.* 28:281–283.
CONN, H. J.
 1933. History of staining. Commission on Standardization of Biological Stains. Geneva, N.Y.
———.
 1953. Biological Stains. 6th ed. Commission on Standardization of Biological Stains. Geneva, N.Y.
CRAFTS, A. S.
 1931. A technic for demonstrating plasmodesma. *Stain Tech.* 6:127–129.
CROWELL, IVAN H.
 1930. Cutting microscopic sections of wood without previous treatment in hydrofluoric acid. *Stain Tech.* 5:149–150.
DAVENPORT, H. A., and R. L. SWANK
 1934. Embedding with low viscosity nitrocellulose. *Stain Tech.* 9:137–139.

DAVIS, G. E., and E. L. STOVER
1936. A simple apparatus for the steam method of softening woods for micro-scope sections. *Trans. Ill. Acad. Sci.* 28:87.

DIONNE, L. A., and P. B. SPICER
1958. Staining germinating pollen and pollen tubes. *Stain Tech.* 33:15–17.

DUFRENOY, J.
1935. A method of imbedding plant tissues without dehydration. *Science* 82:335–336.

EAMES, ARTHUR J.
1936. Morphology of vascular plants. McGraw-Hill. New York.

GAGE, S. H.
1941. The microscope. 17th ed. Comstock Publishing Associates. Ithaca, N.Y.

GOURLEY, J. H.
1930. Basic fuchsin for staining vascular bundles. *Stain Tech.* 5:99–100.

GRAY, PETER
1958. Handbook of basic microtechnique. 2nd ed. McGraw-Hill. New York.

———.
1954. The microtomist's formulary and guide. Blakiston. Philadelphia.

HANCE, R. T.
1933. A new paraffin embedding mixture. *Science* 77:353.

HYLAND, F.
1941. The preparation of stem sections of Woody Herbarium specimens. *Stain Tech.* 16:49–52.

JEFFREY, E. C.
1928. Improved method of softening hard tissues. *Bot. Gaz.* 456–458.

JENSEN, W. A.
1962. Botanical histochemistry, principles and practice. W. H. Freeman and Co. San Francisco.

JOHANSEN, D. A.
1940. Plant microtechnique. McGraw-Hill. New York.

Journal of the Biological Photographic Association
Williams & Wilkins. Baltimore, Md.

KASTEN, FREDERICK H.
1950. Additional Schiff-type reagents for use in cytochemistry. *Stain Tech.* 33:39–45.

KOHL, E. J. and C. M. JAMES
1931. A method for ripening hematoxylin solutions rapidly. *Science* 74:247.

———.
1937. The use of *n*–butyl alcohol in the paraffin method. *Stain Tech.* 12:113–119.

KUHN, GERALDINE D., and E. L. LUTZ
1958. A modified polyester embedding medium for sectioning. *Stain Tech.* 33:1–7.

LEE, A. BOLLES
1950. The microtomist's Vade-Mecum. 11th ed. Blakiston. Philadelphia.

LILLIE, R. D.
1951. Simplification of the manufacture of Schiff reagent for use in histochem-ical procedures. *Stain Tech.* 26:163–165.

———.
1954. Histopathological technic and practical histochemistry. 2nd ed. Blakiston. New York.

LILLIE, RALPH O., WILLIAM F. WINDLE, and CONWAY ZIRKLE
1950. Interim report of the Committee on Histologic Mounting Media. Resin-ous media. *Stain Tech.* 25:1–9.

McLane, Stanley R.
   1951. Higher Polyethylene glycols as a water soluble matrix for sectioning fresh or fixed plant tissues. *Stain Tech.* 26:63–64.

McClintock, Barbara
   1929. A method for making acetocarmin smears permanent. *Stain Tech.* 4:53–56.

McClung, G. E.
   1950. Handbook of microscopial technique. 3rd ed. Paul B. Hoeber, Inc., Harper. New York.

McWhorter, F. P., and E. Weier
   1936. Possible uses of dioxan in botanical microtechnique. *Stain Tech.* 11:107–119.

Maneval, W. E.
   1936. Lacto-phenol preparations. *Stain Tech.* 11:9–11.

Massey, Barbara W.
   1953. Ultra-thin sectioning for electron microscopy. *Stain Tech.* 28:19–26.

Mettler, F. A., C. C. Mettler, and F. C. Strong
   1936. The cellosolve technic. *Stain Tech.* 11:165.

Miller, D. F., and G. W. Blaydes
   1938. Methods and materials for teaching biological sciences. McGraw-Hill. New York.

Newcomer, E. H.
   1938. A procedure for growing, staining and making permanent slides of pollen tubes. *Stain Tech.* 13:89–91.

Newman, S. B., E. Borysko, and M. Swerdlow
   1949. New sectioning techniques of light and electron microscopy. *Science* 110:66–68.

Northern, H. T.
   1936. Histological applications of tannic acid and ferric chloride. *Stain Tech.* 11:23–24.

Photography Through the Microscope
   Eastman Kodak Company. Rochester, N.Y. 1957.

Popham, Richard A.
   1948. Mordanting plant tissues. *Stain Tech.* 23:49–54.

Rawlins, T. E.
   1933. Phytopathological and Botanical Research Methods. John Wiley and Sons, Inc., New York.

Sass, John E.
   1945. Schedules for sectioning maize kernels in paraffin. *Stain Tech.* 20:93–98.
———.
   1947. A photographic apparatus for use with Micro Tessar objectives. *Jour. Bio. Photo. Assoc.* 15:198–200.

Schmidt, Louis
   1937. Photomacrography. *Jour. Bio. Photo. Assoc.* 6:47–61.

Shields, Lora Mangum, and H. L. Dean
   1949. Microtome compression in plant tissue. *Amer. Jour. Bot.* 36:408–416.

Smith, Gilbert M.
   1933. The freshwater algae of the United States. McGraw-Hill. New York.
———.
   1938. Cryptogamic botany. McGraw-Hill. New York. 2 vols.

Smith, Luther
   1947. The acetocarmine smear technic. *Stain Tech.* 22:17–31.

STEERE, W. C.

1931. A new and rapid method for making permanent acetocarmin smears. *Stain Tech.* 6:107–111.

WALLS, G. L.

1932. The hot celloidin technic for animal tissues. *Stain Tech.* 7:135–147.

WEAVER, H. L.

1955. An improved gelatin adhesive for paraffin sections. *Stain Tech.* 30:63–64.

WETMORE, R. H.

1932. The use of celloidin in botanical technic. *Stain Tech.* 7:37–62.

———.

1936. A rapid celloidin method for the rotary microtome. *Stain Tech.* 11:89–93.

WITTLAKE, EUGENE B.

1942. An efficient vacuum apparatus for microtechnic. *Ohio Jour. Science.* 42:65–69.

ZIRKLE, CONWAY

1947. Some synthetic resins in combined fixing, staining and mounting media. *Stain Tech.* 22:87–97.

Index